Holger Flechsig

Molecular Motors

Holger Flechsig

Molecular Motors

Nanoscale engines of the cell and their structurally resolved modeling

Südwestdeutscher Verlag für Hochschulschriften

Impressum/Imprint (nur für Deutschland/only for Germany)
Bibliografische Information der Deutschen Nationalbibliothek: Die Deutsche Nationalbibliothek verzeichnet diese Publikation in der Deutschen Nationalbibliografie; detaillierte bibliografische Daten sind im Internet über http://dnb.d-nb.de abrufbar.
Alle in diesem Buch genannten Marken und Produktnamen unterliegen warenzeichen-, marken- oder patentrechtlichem Schutz bzw. sind Warenzeichen oder eingetragene Warenzeichen der jeweiligen Inhaber. Die Wiedergabe von Marken, Produktnamen, Gebrauchsnamen, Handelsnamen, Warenbezeichnungen u.s.w. in diesem Werk berechtigt auch ohne besondere Kennzeichnung nicht zu der Annahme, dass solche Namen im Sinne der Warenzeichen- und Markenschutzgesetzgebung als frei zu betrachten wären und daher von jedermann benutzt werden dürften.

Coverbild: www.ingimage.com

Verlag: Südwestdeutscher Verlag für Hochschulschriften GmbH & Co. KG
Heinrich-Böcking-Str. 6-8, 66121 Saarbrücken, Deutschland
Telefon +49 681 37 20 271-1, Telefax +49 681 37 20 271-0
Email: info@svh-verlag.de

Approved by: Berlin, TU, Diss., 2011

Herstellung in Deutschland (siehe letzte Seite)
ISBN: 978-3-8381-3408-6

Imprint (only for USA, GB)
Bibliographic information published by the Deutsche Nationalbibliothek: The Deutsche Nationalbibliothek lists this publication in the Deutsche Nationalbibliografie; detailed bibliographic data are available in the Internet at http://dnb.d-nb.de.
Any brand names and product names mentioned in this book are subject to trademark, brand or patent protection and are trademarks or registered trademarks of their respective holders. The use of brand names, product names, common names, trade names, product descriptions etc. even without a particular marking in this works is in no way to be construed to mean that such names may be regarded as unrestricted in respect of trademark and brand protection legislation and could thus be used by anyone.

Cover image: www.ingimage.com

Publisher: Südwestdeutscher Verlag für Hochschulschriften GmbH & Co. KG
Heinrich-Böcking-Str. 6-8, 66121 Saarbrücken, Germany
Phone +49 681 37 20 271-1, Fax +49 681 37 20 271-0
Email: info@svh-verlag.de

Printed in the U.S.A.
Printed in the U.K. by (see last page)
ISBN: 978-3-8381-3408-6

Copyright © 2012 by the author and Südwestdeutscher Verlag für Hochschulschriften GmbH & Co. KG and licensors
All rights reserved. Saarbrücken 2012

Abstract

Motor proteins are complex macromolecules which have evolved through the biological evolution to carry out a variety of functions related to force generation and intracellular transport. Underlying their organized activity are ordered conformational motions induced by binding of ATP molecules and their hydrolysis. Since these cyclic conformational motions are slow, they cannot be reproduced in molecular-dynamics simulations with all-atom models. Therefore, coarse-grained descriptions of reduced complexity are needed.

In this work, a coarse-grained mechanical model, with a protein pictured as a deformable elastic network, has been employed. The focus was on the investigations of helicase proteins, which are molecular motors that translocate in a cell over nucleic acids and unwind their duplex structure. By using the coarse-grained dynamical description for the protein and DNA, and including interactions with ATP molecules, we have successfully followed entire operation cycles of the hepatitis C virus (HCV) helicase, for which a large amount of experimental data is available. Thus, the operation of a real molecular motor could be reproduced - for the first time - in structurally resolved dynamical simulations. Additionally, conformational relaxation dynamics in three other helicases from the same superfamily 2 has been investigated through coarse-grained numerical simulations.

In the last chapter of this book, a different but related problem is addressed. There, we construct and investigate an elastic-network model of a device that can be viewed as a prototype of artificial molecular motors. Similar to myosin motors responsible for force generation in the muscles, the designed machine is able to convert, through a ratchet mechanism, its active cyclic internal motions into a steady net force used to pull a filament. Thermal fluctuations are taken into account and artificial motor operation at different fluctuation levels is discussed.

Contents

1 Introduction   **5**
- 1.1 Proteins and their folded structures 8
- 1.2 Enzymes and motor proteins 10
- 1.3 Helicase motors 12
  - 1.3.1 Helicase classification 13
- 1.4 Experimental investigations of proteins 14
- 1.5 Theoretical modeling of protein dynamics 17

2 Elastic network models   **21**
- 2.1 Proteins as elastic objects 21
- 2.2 Near-equilibrium dynamics 23
- 2.3 Relaxational elastic network model 27
- 2.4 Model extensions and discussion 30

3 Conformational dynamics of Hepatitis C virus helicase   **33**
- 3.1 HCV helicase structure and its elastic network 34
- 3.2 Probing of mechanical properties 36
- 3.3 Visualization of conformational motions 38
- 3.4 Modeling of ATP-dependent operation cycles 40

4 Entire operation cycles of Hepatitis C virus helicase   **47**
- 4.1 Modeling of DNA 47
- 4.2 Translocation mechanism of HCV helicase 51
- 4.3 Duplex DNA unzipping 55

5	Conformational motions in superfamily 2 helicases	**61**
	5.1 *In silico* investigation of superfamily 2 helicases	62
	5.1.1 Hef helicase	66
	5.1.2 Hel308 helicase	67
	5.1.3 XPD helicase	69
	5.2 Functional aspects for helicase operation	71
	5.3 Simulation details	76

6	Models of synthetic protein motors	**79**
	6.1 Introduction	79
	6.2 Design of elastic networks	81
	6.3 Designed network template	85
	6.4 Ligand-induced cyclic operation	87
	6.5 A prototype synthetic protein motor	90
	6.5.1 Interactions with an artificial filament	91
	6.5.2 Motor operation under thermal fluctuations	94
	6.5.3 Filament transport by the motor	96
	6.6 Coordinates of the designed network	98

7	Conclusions	**101**

Bibliography		**105**

Chapter 1

Introduction

Life is possible on the Earth due to the presence of complex molecules such as amino acids that, assembled into larger molecular compounds, can form proteins which represent the principle substance of living organism [1, 2]. Proteins are involved in virtually all processes in the biological cell and exhibit an enormous functional diversity [3, 4]. Their properties have evolved through billions of years of biological evolution with differentiation and specialization in order to perfect a respective operation or develop new functions [5]. When Life is regarded as a manifestation of organized cellular activity, then clearly it is also a direct consequence of the activity of proteins (and their interactions), their evolution and functional promiscuity [6, 7]. On the other hand, anomalies intervening into this perfected system can cause malfunctions, leading to diseases and eventually the knockout of life-essential components [8].

In order to understand the activity of biological cells and the biological processes inside it, it is essential to understand the principles behind the operation of their central players – the proteins.

While the description of proteins as important molecules for the body dates back to the middle of the 19th century [9], their eminent role for living organisms became evident only in the first decades of the 20th century, first in the determination of the complete amino acid sequence of the insulin protein by Frederick Sanger in 1951 [10, 11] and followed by the determination of the first three-dimensional protein structures, those of myoglobin and hemoglobin, by John Cowdery Kendrew in 1958 and Max Perutz in 1963, respectively [12, 13]. These studies, all of them awarded the Nobel prizes later, marked the beginning of extensive research of proteins aimed to understand their structures, functions and their role in the context of cellular activity.

Today, not only the biochemical roles for a great number of proteins are known, but also an enormous wealth of structural data has been collected and its analysis has helped to enlighten our understanding of their functional properties [3, 14]. Furthermore, in the last decades large progress has been achieved in the design of elaborate experimental methods to dynamically observe protein dynamics, aiming to watch them in action, and actively manipulate them to probe their functional properties [15, 16].

Proteins have been categorized into many groups based on the principles of the operation they are carrying out. Of particular interest for dynamical processes of the cell are the motor proteins. They belong to the larger group of enzymes and are able to convert chemical energy into force-producing mechanical motions [17]. This ability makes them the working horses of the cell being able to contribute to many of intracellular processes [18].

The investigation of these important nanoscale machines has attracted the attention of physicists in the last decades and various phenomenological descriptions have been developed to understand the dynamical principles of their operation [19, 20]. The atomistic modeling of protein machines encounters serious difficulties because many biological relevant motions are slow. Because of this, approximate models of intermediate complexity, bridging the gap between purely phenomenological descriptions and models of atomistic detail, have become increasingly popular. Despite their approximate nature, such mid-level descriptions are able to capture functional aspects of conformational motions in motor proteins within their operation cycles remarkably well [21].

The main subject of this Thesis is the application of approximate descriptions to understand dynamical properties of protein machines. While dealing with the systems that are of interest for many fields of science, from chemistry and molecular biology to virology, the primary focus of this work is to study protein machines as physical objects, by using a dynamical physical model. Our findings will however be also discussed with respect to their biological relevance.

CHAPTER 1. INTRODUCTION

This first chapter shall provide a brief introduction to the field of protein research. The functional importance of proteins when acting as enzymes and motors is outlined together with a short survey on the experimental techniques used to investigate structural properties and functional motions. This chapter ends with an introduction into theoretical modeling of the dynamics of motor proteins, explaining the necessity of approximate descriptions in order to understand functional motions appropriately.

In the second chapter, one such approach, viewing the proteins as deformable elastic objects, will be further introduced. Despite strong simplifications, the elastic network model (ENM) has been found to predict functional conformational motions in proteins remarkably well (see e.g. [22]). The approximations upon which the model is based and recent popular implementations are explained. The focus in this chapter will be, however, placed on discussing a particular model used in this thesis – the relaxational elastic network model.

To a large extend, this thesis is devoted to the structurally resolved modeling of full operation cycles of the hepatitis C virus (HCV) helicase motor protein. The investigations of the HCV helicase can be divided into two parts. At the first stage, following the construction of the elastic network, we have probed dynamical properties of this protein. This procedure and the results found by us are described in Chapter 3. Further on we have considered an extension of the model that includes interactions with ATP molecules and the DNA. This is described in the fourth chapter, with the emphasis put on modeling of complete operation cycles of the HCV helicase including its unzipping activity of duplex DNA.

Many different helicase proteins share structural similarities and have related sequential signatures. It has therefore been conjectured that also their ATP-dependent activity may be similar and they can share common modes of operation. We have considered three motor proteins from the major helicase superfamily 2 and used relaxational elastic network models to probe their mechanical properties and to study conformational dynamics that may underly the processing of nucleic substrates in these proteins. This work in reported in Chapter 5.

The dynamical systems investigated in Chapter 6 are of different nature. Theoretical investigations of real biological motor proteins are complicated and computationally

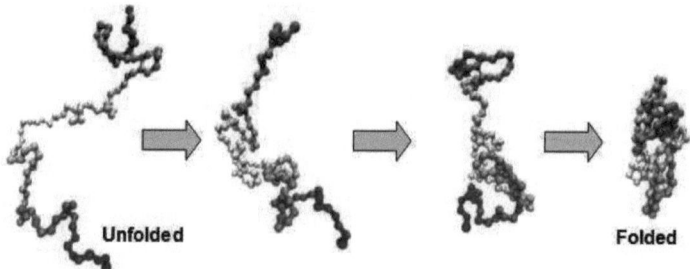

Figure 1.1: **Protein folding (schematic representation).** Under the folding process, the macromolecular shape changes from a chain-like structure (left) to the compact, characteristic native protein structure (right) while crossing complicated intermediate shapes.

extensive. It is therefore beneficial to work instead with the theoretical models dealing with artificially constructed analogs of real biological objects. In the past elastic network descriptions have been used to design networks that are able to perform ligand-induced cyclic conformational changes resembling dynamical properties found in real protein machines. Using the same methods, we have designed model elastic-network machines able to convert ligand-induced internal motions into steady translational motion of a filament. The construction of a model artificial motor and investigation of its behavior in the presence of thermal fluctuations are performed in Chapter 5.

In the last chapter, the results of this work are summarized and the perspectives for further investigations are discussed.

1.1 Proteins and their folded structures

Proteins are encoded by genes and are produced through a process of gene expression [4, 23]. Initially, a protein is just a chain of amino acids. In the chain, attractive and repulsive interactions between the amino acids are however present, rendering a very unstable system. As a consequence, this initial chain undergoes a process of folding towards a three-dimensional compact native structure, unique for each specific amino acid sequence (see Fig. 1.1 [1]) [24, 25, 26]. This folded structure is usually referred to

---
[1] Figure has been taken from the internet.

CHAPTER 1. INTRODUCTION

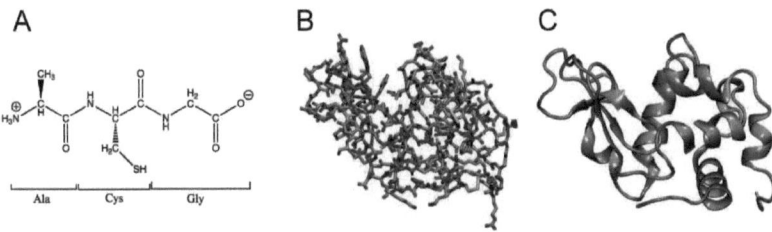

Figure 1.2: **Structural representation of proteins** (A) The chemical structure of a small peptide (i.e. a short protein fragment) with three amino acids Alanine, Cysteine and Glycine. (B) and (C) The hen egg-white lysozyme protein (PDB ID 1LYS) in two common structural representations. In (B) the bonds between atoms are displayed as color-coded sticks. The ribbon representation (C) neglects the amino-acid side-chains but shows common structural motifs, such as helices (the α helices), extended arrows (the β sheets), or thin tubes (flexible loops).

as the equilibrium conformation of a protein. It corresponds to the minimum of energy, with a balance between repulsive and attractive interactions. It should be said that the process of protein folding is highly complicated and its investigations constitute an entire field of research with many open questions [27, 28, 29]. Originated as a problem in molecular biology, protein folding has been however also treated as a biophysical problem and reduced descriptions, employing methods from statistical mechanics, have been developed and extensively studied (see [30, 31] and the review [32]).

The most remarkable property of proteins is probably their evolvability, i.e. their ability to make use of mutations in the genes, and score (or evaluate) them according to the evolutionary pressure, to either adapt perfectly to a particular functional role or develop new folds with different functions [5, 25]. Thus, they are responsible of biological diversity (e.g. [7, 33]). In a fact, their functional variety makes proteins the most ubiquitous molecules in biological systems. For almost every task that needs to be carried out in the cell, some proteins are employed [2, 4].

These macromolecules can operate as structural proteins, able to assemble into filaments and thus offering mechanical support in the cell (actin and tubulin networks). Many hormones that control physiological functions in living organisms are proteins, such as the insulin that regulates glucose concentration in the blood or the growth hormone that can stimulate cell growth and regeneration. In signaling and receptor proteins, the functional emphasis is on the detection of chemical signals and their transduction between the cells or transmission to some processing machinery inside the cell. To display proteins, different representations can be employed, as illustrated in Fig. 1.2.

1.2 Enzymes and motor proteins

An important class of proteins are enzymes [34, 35, 36]. These molecules are able to bind a chemical ligand (the substrate) at a specific site in the protein. This initiates the enzyme's turnover cycle which is to convert the substrate into one or more other chemicals (the products). The cycle can be repeated again and again, so that a certain chemical reaction is continuously carried out [37, 38, 39]. Thus enzymes play the role of catalysts. Their ability to control the chemical reactions makes them essential ingredients of the cell.

Enzymes exhibit high specificity, i.e. they are selective with respect to the chemicals they can bind and thus the chemical reaction they are catalyzing. The strategies used by enzymes to facilitate chemical reactions are manifold and their mechanisms have been extensively studied in terms of classical chemical kinetics, the Michaelis-Menten kinetics [38, 40, 41]. The molecular mechanism by which enzymes are able to perform their operation are studied with classical biochemical methods such as nuclear magnetic resonance or mass spectroscopy [3, 4]. A structurally resolved dynamical monitoring of enzyme operation is however still lacking.

For the activity of the cells, not only the regulation of chemical reactions is of importance. A fundamental role is also played by dynamical processes that involve active transport of molecules through the cell [3, 4, 42]. Such processes need to be carried out in an organized way and indeed there is a special class of proteins that is capable of offering such services – the motor proteins.

Motor proteins are enzymes that catalyze a specific chemical reaction, namely the hydrolysis of a nucleoside triphosphate (usually ATP) upon binding it [43]. They however exhibit remarkable features which make them special as compared to other enzymes. First, they are able to use chemical energy gained from the chemical reaction

CHAPTER 1. INTRODUCTION

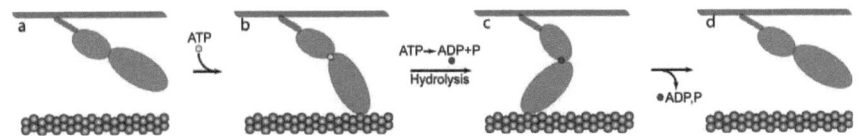

Figure 1.3: **Schematics of an ATP-dependent principle motor cycle.** In its equilibrium conformation (a), the motor is not able to bind to the filament. Binding of an ATP molecule can induce conformational motions such that the binding affinity to the filament becomes very high and a connection can be established (b). Under the hydrolysis reaction large-scale motions inside the motor can be translated into directional motions of the filament – the motor is in the force-producing phase. After this, the motor has still the chemical products bound (c). It is able to return to its free conformation (d) upon their release while having a low affinity for binding of the filament. Under the cyclic operation, the filament became effectively transported by the motor.

(the hydrolysis) and convert it into large-amplitude internal mechanical motions while preserving their folded structure. Second, they can bind to other proteins and use such internal conformational changes to produce a force acting on them [17, 42, 44, 45].

The functioning of motor proteins is based on the repetition of elementary operation cycles. Each such cycle is initiated by binding of an ATP molecule to a specific binding site, a process under which the protein structure can be substantially changed. After that, ATP is hydrolyzed and the gained chemical energy is used to induce further conformational changes. At this stage, the chemical products (ADP and a phosphate) can be released from the protein, allowing it to return to its original equilibrium conformation. This resets the initial motor state, and the next cycle can be initiated by binding of another ATP molecule. An schematic illustration of the operation cycle of a molecular motor is shown in Fig. 1.3.

Motor proteins can be viewed as the working horses of the cell, capable of carrying out a large variety of functions. They are able to assist in the folding of other proteins, make contraction of muscles possible, mediate transmembrane exchange of ions or synthesize polymers [46, 47, 48, 49], etc. Such motor proteins as kinesin and dynein serve as transporters of molecular cargo through the cell. They walk along microtubules. In these molecular motors, cyclic internal conformational changes induced by binding of ATP molecules and its subsequent hydrolysis are translated into directional motions of the motor along the microtubular track [42, 43].

Bruce Alberts, famous for his research in molecular biology, has commented in 1998 on the role of motor proteins for cellular activity in his review paper [18]. He has said that the operation of a cell can be viewed as a result of organized activity of protein machines which can be compared with macroscopic machines invented by humans. Man-made machines are engineered in such a way, that the parts they are containing move in a coordinated fashion in order to execute a certain macroscopic task. Similar to this the execution of microscopic function by a protein machine can be seen as a consequence of ordered and concerted conformational changes in this macromolecule.

1.3 Helicase motors

The genetic information of a living organism is stored in nucleic acids, the DNA and RNA molecules, which are often forming duplex structures [3, 4]. In a variety of processes it is however needed to temporarily separate the duplex into its single-strand components in order to access the sequence of bases. These processes include replication, recombination and repair of nucleic acids. They employ an important class of motor proteins, the helicases [50, 51]. While helicases can also be involved in other kinds of DNA and RNA manipulations, their principal role is to separate duplex nucleic acids into their single-stranded forms [52, 53]. To perform this operation, helicases use cyclic conformational changes induced by binding and hydrolysis of ATP molecules, to translocate along nucleic acids and unwind their duplex structure.

In viruses, helicases are an essential part of the molecular replication machinery [54, 55]. A virus that needs to copy its genome, i.e. its viral DNA or RNA, should employ two main processes. First, the genetic information should be made easily accessible, a task that is executed by the viral helicase. Subsequently, proteins from another class, the polymerases, read out the genetic information and process it to synthesize new viral DNA or RNA [58]. To prevent multiplication of a virus, it might therefore be enough to stop functioning of its helicases. Indeed, there are important antiviral therapies which are based on the drugs that attack helicase motors [54, 55].

Generally, molecular processes that involve helicase operation are diverse and so are the requirements on their operation [59, 60, 61]. One can think of various mechanisms that would make separation of duplex nucleic acids possible. However, the molec-

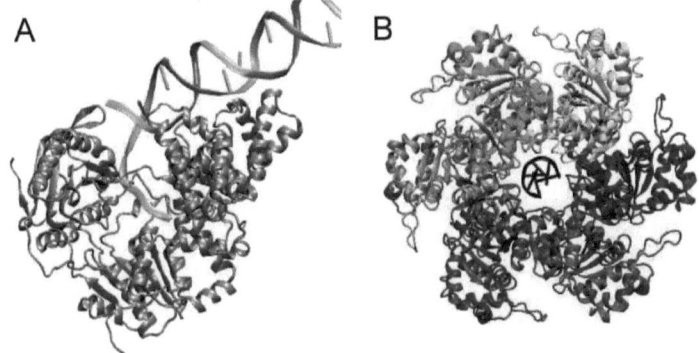

Figure 1.4: **Helicase structures.** (A) The monomeric structure of the UvrD helicase (gray) co-crystallized with duplex DNA (green) [56]. The PDB code is 2IS2. (B) The papillomavirus E1 hexameric helicase with a single DNA strand occupying the central hole cavity [57] (PDB ID 2GXA). The six similar subunits are colored differently and the DNA is shown in black.

ular toolkit provided by nature to perform such operations is actually limited ([53]). Although a large number of different helicases in various organisms exist, they have structural similarities and share common principal sequential fingerprints [62].

1.3.1 Helicase classification

A large number of helicases from different organisms has already been identified, characterized and grouped according to their structural and sequential aspects (see [62] and the review [59]). A distinction between monomeric and oligomeric helicases should be made. While most helicases are functional as monomers, there are also interesting oligomeric forms [63]. For example, in hexameric helicases six copies of the same monomer are arranged in a symmetric ring-like architecture with a hole in the center through which one nucleic acid stranded can be pumped by coordinated movements of the individual subunits. This cooperative dynamics leads to the separation of the complementary strand [57]. In Fig. 1.4, examples of a monomeric and hexameric helicases are given.

Restricting out attention to monomeric helicases, it can be noted that, although the number of such molecules employed by the organisms to process DNA and RNA is large, many of them have a similar folded structure [50, 59].

In the largest superfamily 2, on which the focus has been put in this work, the similarity among its members is the presence of two structurally similar core domains [59]. Besides of the structural aspects, proteins in this superfamily also share a number of conserved residue motifs. The conserved motifs are certain short sequences of amino acids that are always found. Such characteristic sequences are located in different parts of the proteins involved in the processes including ATP and the nucleic acids [59, 64]. Some important motifs essential for binding and hydrolysis of ATP molecules are found in the core domains. They are the Walker A motif (also referred to as motif I), the Walker B motif (referred to as motif II) and the motif VI [59, 62].

The fact that those residue groups are conserved means that they must be essential for the operation and needed to be taken into account in the dynamical modeling of a helicase motor. In the simulations of the hepatitis C virus helicase, which represents the most prominent member of the superfamily 2 helicases, we will exclusively use the conserved residues when interactions with the DNA and ATP will be considered. The same applies to the investigation of the dynamical properties undertaken by us for other helicases from the same superfamily.

1.4 Experimental investigations of proteins

The structure of a protein determines its functional properties and today several methods are in use to resolve it in the experiments [3, 4]. The hemoglobin, the protein that transports oxygen through the blood, and the related myoglobin, were the first proteins whose three-dimensional structure could be determined in 1958 by Perutz and Kendrew – a discovery that was awarded with the Nobel prize for chemistry [12, 13]. The technique used in such experiments was the X-ray diffraction analysis. This technique requires to first crystallize the protein. The crystal is then exposed to X-ray radiation which generates a pattern of reflections that encodes the spacing of atoms in the crystal lattice. This pattern of spots can be analyzed to gain information about the spatial coordination of atoms and thus to generate a three-dimensional structure of the protein [65, 66]. Although, at present, this is the standard method to determine the structure of proteins, it has serious drawbacks. Not only the fabrication of the protein crystal may

CHAPTER 1. INTRODUCTION

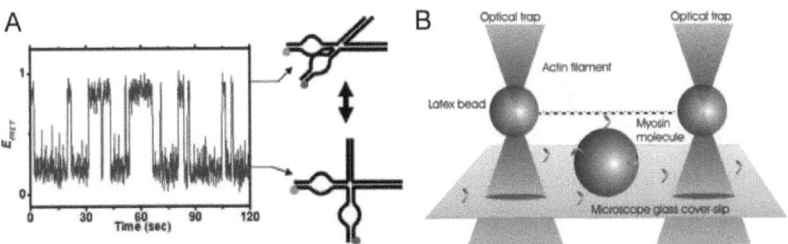

Figure 1.5: **Single-molecule experiments.** (A) Fluorescent resonant energy transfer (FRET) method. The time series of the FRET signal allows to detect functional motions inside the protein. This particular example refers to the folding and unfolding dynamics of donor and acceptor-labeled RNA. (B) An optical tweezer setup to study interactions between myosin motors and actin filaments [3].

be complicated since some proteins can be hardly crystallized, but, moreover, the protein structure can be perturbed during this process, so that flexible parts of the protein cannot be determined within good resolution.

Some of these difficulties can be circumvented by the nuclear magnetic resonance (NMR) spectroscopy method, allowing to carry out structure analysis directly in the solution and being able to shed light on the dynamical aspects, such as the conformational changes [67, 68]. This method relies on nuclear spin properties of the atoms and the alignment of their magnetic dipoles when strong magnetic fields are applied to the solution. This allows to measure an energy absorption spectrum which reflects the identities of the atoms and their chemical environment. The spectrum can be further processed to reconstruct the three-dimensional structure of the protein.

Both techniques have been widely applied in the past to determine the three- dimensional structure of biological macromolecules [25]. The knowledge of protein structures has largely contributed to the understanding of physical properties of proteins and their functionality, the structure-function relation [3, 4]. The great amount of information is stored in the Protein Data Bank[2] (PDB) which can be easily accessed through the internet. Currently this archive contains more than 75,000 different protein structures.

For many proteins, not only the native unliganded structure is experimentally available. It is possible to find even structures that capture the protein with various chemical ligands bound to it, thus providing snapshots of the operation cycles. Such data can

---

[2] www.pdb.org
[3] (A) taken from the Ha lab (http://bio.physics.illinois.edu/), (B) image courtesy of Alexandre Lewalle

1.4. EXPERIMENTAL INVESTIGATIONS OF PROTEINS

yield valuable insights into the conformational changes accompanying the protein operation. To work with such data, Gerstein and Krebs have developed the Database of Macromolecular Movements [4]. This database stores the structures and allows to see motions between the structural states created by using the interpolation algorithms such as morphing [69, 70]. Thus, basic domain movements in proteins were identified and classified according to the principal types of molecular motions observed [71].

To understand the operation mechanisms of proteins, in particular of motor proteins, the best one can imagine is to watch them in action with an atomic spatio-temporal resolution. While this is still a fiction, since no microscope is able to offer such a service, large experimental progress has been nonetheless made in the last years [15, 16].

A state-of-the-art setup to follow the time course of conformational motions in proteins is provided by the fluorescent resonant energy transfer (FRET) method [72, 73]. Its mechanism is based on energy transfer between two marker molecules, the donor and acceptor, which strongly depends on the spatial distance [74, 75]. When such markers are attached at two different domains in the protein (and their distance is not too large, i.e. < 10 nm), this method allows to detect dynamical changes in the protein conformation [76].

In addition to observing dynamical behavior, it is furthermore possible to actively probe the dynamics of motor proteins in the experiments. For such purposes optical tweezer setups or the atomic force microscopy (AFM) are employed [77, 78, 79]. Optical tweezers use highly focused laser beams to trap proteins that prior to this have been manipulated with dielectric beads. In this way the protein can be exposed to various directed forces, to probe its mechanical properties [80]. In AFM configurations, proteins need first to be deposited on surfaces, in order to further test their mechanical properties with a microscopic tip [81, 82]. In Fig. 1.5, examples of single-molecule experiments are given.

Finally, two fascinating recent studies should be mentioned. In Tokyo, Ando with coworkers have reported the investigations of the dynamics of two important motor proteins. Using high-speed atomic force microscopy, they could directly visualize the

---
[4] http://molmovdb.org

myosin V protein as it walks along the actin filament tracks, producing a video with high resolution [83]. Proceeding further, a similar technique has been used to observe dynamical changes in the F1-ATPase motor, directly visualizing the rotational dynamics in its oligomeric structure [84].

Although time-resolved single molecule experiments can characterize conformational changes in proteins and therefore provide insights into their functional properties, the present situation is that the investigations of protein motors still rely to a large extend on the static structural information. In the experiments, it is not possible to dynamically follow the operation cycles in a structurally resolved manner.

1.5 Theoretical modeling of protein dynamics

Internal motions in motor proteins occur on the timescales ranging from picoseconds for amino acid side-chain fluctuations to milliseconds or even seconds for conformational changes within the turnover cycles. The most accurate description of protein dynamics is provided by all-atom molecular dynamics (MD) simulations [85]. In such approaches, all particles of the protein and all interactions between them are included into a computation [86, 87, 88]. The interatomic potentials are complicated and aim to describe the interactions in a detailed way. They contain various variables such as the bond length, the bond angles and the dihedral angles.

Both the large number of particles and the use of the full potentials in MD simulations lead to heavy computational expenses. The time steps needed in the integration of the equations of motion for all particles are typically femto-seconds and therefore the accessible timescale covered by MD simulations is typically at the order of nanoseconds [89, 90, 91]. The application of such methods is therefore restricted to molecular processes that occur on short timescales, often spatially localized in some active regions of the protein, such as binding of a ligand or dissociation of reaction products.

Slow internal motions, that are crucial for biological operation of proteins, cannot be followed in such simulations. The modeling of entire operation cycles in typical motor proteins, with the timescales ranging from tens of milliseconds to even seconds,

1.5. THEORETICAL MODELING OF PROTEIN DYNAMICS

lies far beyond the feasibility of all-atom MD simulations, even if the latest computer architectures are used. Only for a very small protein, an extremely long MD-simulation reaching 1 millisecond has been reported [92].

Thus, atomically resolved simulations of molecular processes, underlying the operation of protein motors, is not possible and some approximations need to be applied.

As an alternative to detailed molecular simulations, largely simplified descriptions in terms of mechanical ratchets or stochastic oscillators have been used in the studies of the dynamics of motor proteins (see e.g. [19, 20, 93]). In these purely phenomenological models, the complexity of the motor dynamics is reduced to effective motions along a single mechanical coordinate. Such studies of oversimplified models may still provide valuable general information As an example, a ratchet model of the molecular motors is briefly outlined below.

In a typical ratchet model, the structural complexity of a motor protein is neglected and it is viewed as a point particle that can move in a one-dimensional potential. This potential is chosen in the form of an asymmetric sawtooth potential (the ratchet potential), to roughly imitate the interactions between the protein motor and the molecular track along which it is moving. The motor is assumed to have two internal states. In its non-excited equilibrium state, the motor rests at the bottom of the local potential well, unable to steadily move along the track even in the presence of thermal fluctuations. To roughly model the ATP-dependent operation of the motor, it is further assumed that the interactions with ATP can lift the energy of the protein to a level above the barrier of the ratchet potential. When the cycle ends and the motor returns to its non-excited state, it can find itself resting in the neighboring well in the energy landscape, so that the translocation by one step along the track has taken place. This outline corresponds to the simplest ratchet modeling. Other ratchet descriptions, including more complex dynamical aspects, are also available [94, 95, 96].

The two presented approaches, i.e. the full MD-simulations and the ratchet theories, can be seen as two extremes. While, due to their high complexity, the all-atom descriptions fail to capture the biologically essential slow dynamics in motor proteins, the ratchet models only provide an oversimplified picture of their operation mechanism, completely lacking the structural information and thus missing dynamical changes in the proteins. It is desirable to make use of models that can provide mid-level descriptions of protein dynamics, filling the gap between the above mentioned extreme approaches. Such approaches should allow us to follow conformational motions under structural resolution and therefore enable us to analyze in detail the functional dynamics in motor proteins. Such mid-level models should represent certain approximations/simplifications to the full molecular dynamics. Generally, the approximations can be applied either at the structural level or regarding the interaction potentials, and the degree of the reduction of complexity can be varied [21, 97, 98, 99].

Structural coarse-graining aims to drastically reduce the number of variables that enter the calculations. Rather than taking into account each atom of the protein explicitly, whole atomic groups (residues) can be treated as single sites. In the so called single-site per amino acid method, each each amino-acid residue in the protein is replaced by one site which is represented by a point-like particle. Commonly, this particle is placed at the position of the α-carbon atom of the main-chain of the respective amino acid (the C-α representation). Since the side-chain packing of the different amino acids is then neglected, the position of the single sites can alternatively be determined as the geometrical center of the heavy side-chain atoms of the given amino-acid (the C-μ representation). In a more refined structural approximation, two sites can be used for the representation of each amino acid, one replacing its main-chain atoms and the other replacing its heavy side-chain atoms. A nice review focusing on coarse-graining in proteins has been written by Tozzini [98].

In the approximate descriptions, simplifications regarding the interaction potentials are intended to reduce the dynamical degrees of freedom that enter the computation. In the past, Go-like potentials, based on ideas of a work by Go et al on the theory of protein folding [100], have been much used as a possible simplification of the residue potential fields. The Go-like models distinguish between interactions between the consecutive residues inside the backbone chain of the protein, the so called non-native contacts, and the interactions between the residues that are present due to the compact folded structure of the protein, the native interactions [101]. In such models, both types of interactions are simplified, so that only the distance between interacting residues enters the potentials. Non-native backbone contacts are usually described by bounded attractive potentials, taking into account the backbone stiffness, while native contacts are modeled by phenomenological Lennard-Jones type potentials, effective only if the distance between two residues is shorter than some interaction radius.

Chapter 2

Elastic network models

As we have mentioned at the end of the previous chapter, descriptions of intermediate complexity based on approximations are needed in order to cover biologically relevant slow conformational motions in proteins in dynamical simulations. The approximations can be manifold and are typically carried out by simplifying the structure of a protein and reducing the level of complexity of internal interactions. A remarkably simple dynamical descriptions treating the protein as an elastic object is presented in this chapter. We will first describe the approximations this model is based on and present its widespread usage together with a discussion on their limitations. Finally, the relaxational elastic network model, which is used by us to understand the dynamics of motor proteins, will be introduced.

2.1 Proteins as elastic objects

In the framework of elastic network models (ENM) the protein is described as a deformable elastic object. This idea has been introduced by Tirion in her seminal paper in 1996 [22]. Essentially she has shown that instead of sophisticated multi-parameter potentials used for modeling of interactions between atoms in a protein, a simple single-parameter Hookean type potential is appropriate enough to reproduce well the complex deformation pattern induced by thermal fluctuations in proteins. This work has marked a milestone in the modeling of protein dynamics since it allows to study dynamical properties of relatively large macromolecular systems with manageable computational expense. Although in the original description by Tirion the harmonic potentials were

2.1. PROTEINS AS ELASTIC OBJECTS

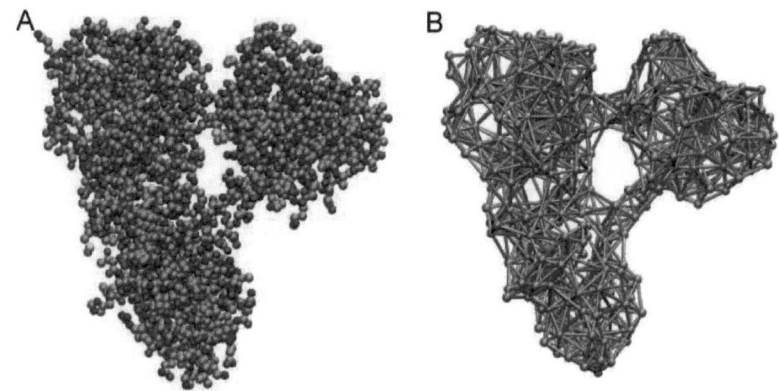

Figure 2.1: **Coarse-graining of proteins.** The structure of the HCV helicase motor that plays a central role in our investigations is shown (PDB code 1HEI, chain A). In (A) a representation showing all presents 3328 atoms is chosen. In (B) only the 443 Cα-atoms, which replace entire amino acids upon structural coarse-graining, are shown. A prescribed interaction radius of 8Å gives rise to 2235 links connecting these particles.

used to mediate interactions between atoms in the protein, the same potentials can be however also applied to a protein model that implies further coarsening of the protein structure as evidenced by the work of Bahar and Hinsen [102, 103]. There, it has been shown that one single interactions site per amino acid residue is sufficient to describe thermal fluctuation motions. In most elastic network descriptions, structural coarsening is implemented today.

Thus, within the simplifications of the elastic network approximation, a protein can be viewed as a network of particles connected by elastic springs.

We will now describe in more detail how the elastic network of a protein is constructed. The model assumes the equilibrium conformation of the protein to be known from experiments (see 1.1.4). Taking this data, each amino acid residue is replaced by a single point-like particle which is placed at the position of the α-carbon atom of the respective amino acid. By carrying out this structural coarsening, the number of particles joining the dynamical simulations is generally drastically reduced (see Fig. 2.1). Then, effective interactions between the particles are introduced by assuming that only those which are close enough one to another should 'feel each other'. This is done by connecting those pairs of particles with a spring that have a distance shorter than a prescribed cutoff-length (also referred to as the interaction radius in the thesis). In the protein network, all particles are assumed to be identical and also the springs are chosen to be of the same kind, meaning that they have a common stiffness. This setup can be viewed as the 'simplest', most basic version of elastic network models [104, 105, 106]. There exist however several variations where different interaction strengths have been also considered. We refer to section (2.4) for a discussion of such issues.

2.2 Near-equilibrium dynamics

The elastic network model provides a purely mechanical description of protein dynamics. While the operation of proteins apparently involves both mechanical and chemical aspects, the ability to cover the latter has become lost due to the approximations; elastic network models are inherently blind against chemical details of the protein activity.

As we have mentioned in the last section, the elastic network model requires the knowledge of the equilibrium structure of the protein. The corresponding elastic network can be constructed following the aforementioned recipe. The dynamics of this network is specified below.

First we introduce some notations. The number of network particles will be denoted by N. The connectivity of the network, i.e. the information which particle is connected to another one, is stored in the symmetric $N \times N$ adjacency matrix $\mathbf{A}$ with entries $A_{ij} = 1$, if the particle with index i is connected by a spring to the particle with index j, or $A_{ij} = 0$, else. The spatial coordinates of network particles as extracted from the known equilibrium structure are stored in the vectors $\vec{R}_i^{(0)} = (x_i^{(0)}, y_i^{(0)}, z_i^{(0)})^T$.

Let us now consider the energy function U of the system. The energy can be simply written as the sum of all contributions of network springs, each represented by the respective harmonic potential. It reads as

$$U = \frac{\kappa}{2} \sum_{i<j}^{N} A_{ij}(d_{ij} - d_{ij}^{(0)})^2. \qquad (2.1)$$

Here $d_{ij} = |\vec{R}_i - \vec{R}_j|$ are the lengths of the springs in some deformed network conforma-

2.2. NEAR-EQUILIBRIUM DYNAMICS

tion and $d_{ij}^{(0)} = |\vec{R}_i^{(0)} - \vec{R}_j^{(0)}|$ are their lengths in the respective equilibrium conformation. As mentioned before, the links are assigned the same stiffness constant κ. Their natural lengths $d_{ij}^{(0)}$ are however individual.

In general, the dynamics of a mechanical system of N particles follows the Newtonian equations of motion and can be written in the form

$$M\ddot{\vec{R}} = -\vec{\nabla}U(\vec{R}), \qquad (2.2)$$

with the $3N$-dimensional vector $\vec{R} = (R_1, \ldots, R_{3N})^T = (x_1, y_1, z_1, \ldots, x_N, y_N, z_N)^T$. The $3N$-dimensional gradient vector on the right side is $\vec{\nabla} = (\partial/\partial x_1, \partial/\partial y_1, \partial/\partial z_1, \ldots, \partial/\partial x_N, \partial/\partial y_N, \partial/\partial z_N)^T$.

Although in the protein network each contribution by a single spring is described by individual harmonic interactions, the overall potential depends on all $3N$ spatial coordinates $x_1, y_1, z_1, \ldots, x_N, y_N, z_N$ and is generally anharmonic. As we will see in the following, the dynamical description of the system can be remarkably simplified when the near-equilibrium behavior is considered.

We are considering now only small deviations of the particle positions from their equilibrium positions $\vec{r}_i = \vec{R}_i - \vec{R}_i^{(0)}$. Then, the energy function U can be expanded around the equilibrium conformation of the system according to Taylor, which yields

$$U(\vec{R}^{(0)} + \vec{r}) = U(\vec{R}^{(0)}) + \sum_{k=1}^{3N} \frac{\partial U(\vec{R})}{\partial R_k}\bigg|_{\vec{R}=\vec{R}^{(0)}} r_k + \frac{1}{2} \sum_{k=1}^{3N} \sum_{l=1}^{3N} \frac{\partial^2 U(\vec{R})}{\partial R_k \partial R_l}\bigg|_{\vec{R}=\vec{R}^{(0)}} r_k r_l + \mathcal{O}(r_i^3) \quad (2.3)$$

By construction of the network, the first term, being the elastic energy of the network in its equilibrium conformation, vanishes (see equ. (2.1)). The second term corresponds to internal forces in the equilibrium conformation of the network which also vanish since all particles are at their equilibrium positions.

This means that the energy function considered around the equilibrium conformation of the system simplifies to

$$U(\vec{R}^{(0)} + \vec{r}) = \frac{1}{2}\vec{r}^T H \vec{r}, \qquad (2.4)$$

where the $3N \times 3N$ so-called Hessian matrix H with entries $H_{kl} = \frac{\partial^2 U(\vec{R})}{\partial R_k \partial R_l}\bigg|_{\vec{R}=\vec{R}^{(0)}}$ has

CHAPTER 2. ELASTIC NETWORK MODELS

been introduced.

After the approximation of the potential function, the components of the forces simplify since we have $\partial U(\vec{R})/\partial R_k = \partial U(\vec{R}^{(0)} + \vec{r})/\partial r_k = (H\vec{r})_k$. The Newtonian equations (2.2) become therefore

$$\ddot{\vec{r}} = -H\vec{r}. \tag{2.5}$$

Here, we have assumed the same common mass for all N particles and set is equal to unity.

Now, the worth of the near-equilibrium approximation of the potential function U becomes apparent: The equations of motion of the N particle system (2.5) are well-known vibration equations that can be solved analytically. Its solutions are determined by the eigenvalue problem

$$H\vec{u}_k = \omega_k^2 \vec{u}_k. \tag{2.6}$$

Diagonalization of the Hessian matrix H directly yields a set of eigenvalues ω_k and eigenfunctions $\vec{u}_k$.

The near-equilibrium approximation explained above is applicable to any dynamical system whose dynamics can be described by a potential function that obeys a global minimum. For the particular energy function of the elastic protein network (2.1), the Hessian matrix contains 3×3 block matrices which read as $H_{i,j}^{\alpha,\beta} = \kappa u_{ij,\alpha}^{(0)} u_{ij,\beta}^{(0)}$. Here, i,j are particle indices, $\alpha, \beta = x,y,z$ and $u_{ij} = (\vec{R}_i^{(0)} - \vec{R}_j^{(0)})/d_{ij}^{(0)}$.

As we have learned now, close to the equilibrium the dynamics of the elastic networks can be described in terms of natural vibration frequencies ω_k and corresponding normal modes $\vec{u}_k$. The analysis of the equations (2.6) is referred to as the normal mode analysis (NMA).

The recipe for the normal mode analysis is actually straightforward. Given the potential function of a system, one needs to calculate the Hessian matrix and perform its diagonalization to access the frequencies and modes that determine the near-equilibrium dynamics. It should be noted that the NMA is a well-established approach to study the dynamics of macromolecular systems, and, even before the simplified har-

2.2. NEAR-EQUILIBRIUM DYNAMICS

monic potentials have been introduced by Tirion, such methods have been used to investigate conformational motions in proteins [107, 108, 109, 110].

It is surprising how many results can be obtained by using such purely mechanical model based on gross simplifications. After Tirion has first shown that fluctuations in proteins can be described with the normal modes of an elastic-network model, such models became increasingly popular and their applicability has been tested in many studies (e.g. [104, 105, 106]).

It has been shown, for instance, that for some proteins the conformational changes between two crystallographically known conformations show remarkable overlaps with one or several low-frequency normal modes of the corresponding elastic network [111]. Furthermore, an important work by Zheng and Doniach has revealed that ligand-induced conformational motions in such important motor proteins as myosin or F1-ATPase agree well with the dominant lowest-frequency modes of the elastic networks [112].

These and other studies suggest that conformational motions in proteins may be already described in terms of the collective coordinates corresponding to a few normal modes. It may seem that, to understand the activity of proteins, it might be enough to consider only the normal modes of an the elastic network. The situation is not, however, so simple. Already in the study by Zheng [112], it has been actually shown that the normal mode descriptions is breaking down when conformational motions in the motor protein kinesin are, for example, considered.

It should be remembered that the normal-mode analysis relies on the approximation of the energy function around the equilibrium conformation of the elastic network, which should hold only for small deviations of particles from their equilibrium positions, i.e. only small-amplitude fluctuation dynamics can thus be described. The fact that even large-amplitude functional conformational motions in some proteins, where the protein system should be definitely operating under off-equilibrium conditions, can sometimes be successfully described by this approximation is remarkable. However, when a particular protein and its representation as an elastic network are considered, it is generally not known in what range the normal mode approximation would be valid. One cannot simply assume that large-scale motions in proteins are described in terms of the near-equilibrium properties of its elastic network; the validity of normal-mode analysis is always under question.

The applicability of the normal-mode approximation has been discussed in the literature, particularly in an important contribution by Jernigan [113]. In this publication, he addressed the question to what extend normal modes of elastic networks are capable of covering large-scale protein motions. The results obtained in this study suggest that the normal-mode description generally works better, when collective motions in proteins are considered, and it fails when the degree of collectivity is low.

Because of the known limitations of the normal-mode description, it may be more convenient to go beyond the NMA approximation and consider the nonlinear dynamics of the elastic networks.

2.3 Relaxational elastic network model

We start again with the potential of the elastic network given by equ. (2.1). Under the assumption that hydrodynamic interactions of the protein with its surrounding aqueous solution can be neglected and that thermal fluctuations are not taken into account, the dynamics of a purely mechanical elastic network model can be expressed in terms of a set of differential Newtonian equations, each of them describing the motion of one network particle. Using previous notations, the equation for particle i reads as

$$m\ddot{\vec{R}}_i + \gamma \dot{\vec{R}}_i = -\frac{\partial}{\partial \vec{R}_i} U. \qquad (2.7)$$

In this equation, the inertial forces and a friction term accounting for interactions with the solvent (assuming the same friction coefficient γ for all particles) are present. The derivative of the elastic energy U accounts for the internal elastic forces. The characteristic ratio between the inertial and the frictional forces depends sensitively on the system parameters and determines the dynamical behavior of the system (see [114]). When motions of proteins in their surrounding media are considered, viscous friction forces should dominate the dynamics [114]. Indeed, it has been previously demonstrated through direct molecular dynamics simulations that the inertial effects can be neglected already for motions with the timescales of tens of picoseconds [115]. Since much slower conformational motions with the timescales of milliseconds will be considered, the frictional forces always would dominate the dynamics and the inertial forces can be neglected in our investigations. Hence, the equations of motion can be chosen in the overdamped limit, so that the network dynamics becomes purely dissipative.

Neglecting inertial terms and substituting explicit expressions for the elastic energy the following dynamical equations are obtained

$$\dot{\vec{R}}_i = -\Gamma\kappa \sum_{j=1}^{N} A_{ij}(|\vec{R}_i - \vec{R}_j| - |\vec{R}_i^{(0)} - \vec{R}_j^{(0)}|) \frac{\vec{R}_i - \vec{R}_j}{|\vec{R}_i - \vec{R}_j|}, \qquad (2.8)$$

where the particle mobility $\Gamma = 1/\gamma$ is introduced.

Note that elastic forces depend linearly on the elongations of the springs connecting the particles. Nonetheless, the dynamics of the network is still nonlinear, since distances are nonlinear functions of the $3N$ spatial coordinates, i.e. $d_{ij} = |\vec{R}_i - \vec{R}_j| = \sqrt{(x_i - x_j)^2 + (y_i - y_j)^2 + (z_i - z_j)^2}$.

We shall not perform further linearization of the dynamical equations. This set of equations already completely describes the protein dynamics in the considered elastic network approximation.

Elastic objects can undergo large-scale deformations of their global shape while locally deformations still remain small. Indeed, a rubber band can be, for example, stretched very much without local rupture. We can therefore expect that the relaxational elastic network models should be able to describe well the large-amplitude conformational motions in proteins in which local network deformations are still elastic and described by Hooke's law.

Once the elastic network of a particular protein has been determined, the set of dynamical equations (2.8) can be numerically integrated to obtain the positions of particles $\vec{R}_i$ at any time moment. For numerical implementation, standard methods, such as the Euler integration scheme or the Runge-Kutta method, can be employed.

In the simulations described in the following chapters, the rescaled version of the dynamical equations (2.8) is used. Introducing the dimensionless time $\tau = \Gamma\kappa t$, we get

CHAPTER 2. ELASTIC NETWORK MODELS

rid of the parameters Γ and κ, so that the equations take the form:

$$\dot{\vec{R}}_i = -\sum_{j=1}^{N} A_{ij} (|\vec{R}_i - \vec{R}_j| - |\vec{R}_i^{(0)} - \vec{R}_j^{(0)}|) \frac{\vec{R}_i - \vec{R}_j}{|\vec{R}_i - \vec{R}_j|} \qquad (2.9)$$

These equations provide the basis for the investigations carried out in this thesis. As we see, within the framework of the relaxational elastic network model, protein motions are described in terms of processes of conformational relaxation of the corresponding elastic network, going downhill in the energy landscape and tending to reach the equilibrium state.

We should note, that in all simulations performed by us only elastic proteins motions have been considered. Therefore, breaking of network links was not included into the model. However, in principle, the model can be extended to incorporate plastic deformations also [116, 117].

Relaxational elastic network models have previously been used by Togashi and Mikhailov [118] in their study of conformational motions in several typical molecular motors. In a special study by Togashi et al using the same model [119], effects of nonlinearities in the motions of motor proteins have been extensively discussed. It has been shown that relaxation dynamics in the motor protein myosin V is in good agreement with the predictions based on the normal modes whereas such methods fail completely for another molecular motor, the kinesin KIF1A.

The dynamical equations describing relaxational motions of the elastic network only include internal elastic forces. For the upcoming applications, where we plan to access dynamical properties of the proteins by probing the response to initial deformations of its corresponding elastic network, we need to include external forces. On the other side, by allowing such additional forces, it is also possible to include thermal fluctuations into the model, as has been done previously [117, 118, 120]. Introduction of external forces can be performed in a straightforward way, by adding the force $\vec{f}_i$ acting together with the internal forces on the network particle i to the equations of motion (2.9).

2.4 Model extensions and discussion

The fact, that a simple elastic-network model of proteins based on gross simplifications and covering only mechanical aspects of the protein, is so successful is quite remarkable. Today, different aspects of the elastic-network model have been discussed and various extensions have been proposed. Below, some of such work is briefly outlined.

One issue of the elastic network modeling is the choice of the cutoff-length or the interactions radius which is used in the network construction based on the experimental data. This important parameter remains to some extend arbitrary. It cannot be chosen to be too small, because then the networks tend to include weakly connected residue groups or even disconnected components. The interaction radius cannot be also too large, because then many cross-links between the domains become present, preventing large-scale domain motions.

An attempt to get rid of the cutoff-length as a free parameter for the protein interactions has been undertaken by Jernigan [121]. He has proposed the parameter-free elastic network model, where it is assumed that interactions are present between all network particles, but their interaction strengths are described by spring constants that are inversely proportional to the actual distance between them.

In the original model and also in the relaxational ENM used by us, the springs are assumed to have the same common stiffness. In many publications, however, weighted interactions between residue particles have also been considered. The use of non-uniform spring constants depending, for instance, on the distance between residue particles have been proposed and extensively discussed by Hinsen with coworkers [97, 122, 123, 124, 125, 126]. Moreover, there are also some ENM studies where the elastic system is divided into rigid bodies and flexible regions [127].

In original elastic-network models, the dynamics depends only on the elongations of the springs. There are model extensions, where additional dynamical degrees of freedom have been included into the model, such as the bending or torsional angles [128, 129].

The proposed model extensions are usually tested by checking how well they are able to predict thermal fluctuation motions of the residues around the equilibrium conformation in various proteins. The quality of this benchmark prediction is often taken as a justification of a particular improvement. It is however clear that elastic network models can provide only a simplified description of protein dynamics. Like any other model that relies on approximations, it is also inherently limited in its predictive power. This is mainly manifested in the fact that the elastic network models can only describe mechanical aspects of protein motions while important molecular events rely on the chemical details which are not at all included in the elastic network models. Moreover, the Hookean potentials used to describe interactions between protein residues are empirically chosen and not the result from a rigorous derivation. While in the MD simulations the interactions are complicated and can include also molecular details, the empiric harmonic potentials in the elastic-network models do not at all correspond to particular molecular interactions but only account for the effective interactions in a protein.

It is clear that elastic networks do not reproduce functional motions in proteins in all details, and they cannot replace all-atom MD simulations. In this thesis, only the standard relaxational elastic-network model will be used and we explicitly want to take advantage of its simplicity.

Chapter 3

Conformational dynamics of Hepatitis C virus helicase

The hepatitis C virus is responsible for most cases of non-A and non-B hepatitis infections, mainly associated with unscreened blood transfusions and unsafe injection practices [130, 131]. According to the world health organization (WHO), about 3% of the world population are chronically infected with HCV with 3 to 4 million newly infections per year, making the virus a major health concern. About 80% of newly HCV-infected individuals go on to develop chronic infection with long-term complications including liver cirrhosis and liver cancer. Today, there is no vaccine available and since therapies are not sufficiently successful, much research has focused on the proteins that control replication as possible targets for a more effective treatment [54, 55, 132].

To stop viral multiplication, it would be indeed very efficient to inhibit those proteins, that operate at the beginning of the replication process. In this respect, the understanding of dynamical properties of the HCV helicase may provide valuable insights into its operation mechanism and therefore be an important step towards the development of better antiviral therapies.

The HCV helicase is one of the best characterized helicase proteins. Based on the X-ray crystallographic experiments, the first structures have been reported by Yao et al in 1997 [133] and with a bound oligonucleotide in 1998 by Kim et al [134]. Moreover, time-resolved single molecule experiments using an optical tweezer setup [135] and FRET techniques [136] have been undertaken for this protein. Later, it became also possible to co-crystallize HCV helicase with different ATP mimicking molecules and with DNA [137], and its natural substrate RNA [138], respectively.

Previous studies have contributed much to the understanding of functional activity of this protein. Although the available data provides insights into various aspects of functional activity, the suggested operation mechanism of HCV helicase is still hypothesis, not directly confirmed by any experiment.

Our main goal in this work was to use coarse-grained relaxational elastic network models to investigate conformational dynamics in the HCV helicase protein and to extend them in such a way, that eventually we would be able to model its operation cycles including interactions with ATP molecules and the DNA, in structurally resolved dynamical simulations. The presentation of this project is divided into two parts. Below in this chapter, it is explained how the elastic network has been set up using the structural data. Moreover, treating the protein as a deformable elastic object, we describe how we have probed the dynamical properties by studying its response to various structural deformations. Based on the results of these investigations, modeling of interactions with ATP molecules and with DNA strands and, furthermore, the simulations of entire operation cycles of the HCV helicase will be presented in the next Chapter 4. The results described in this and the following chapter have been published in the article [139]. Videos presenting these results can be accessed online via the PNAS webpage.

3.1 HCV helicase structure and its elastic network

The HCV helicase is a 450 amino-acid residue protein whose folded structure consists of three domains that have comparable sizes (see Fig. 3.1). They are arranged in a Y-shape-like structure of dimensions 65Å and 35Å. Two domains in this protein, the so-called motor domains, are structurally similar, sharing the RecA-like fold. The overall protein structure comprises two large clefts. In the first cleft which separates the two motor domains, ATP molecules can arrive and bind to one motor domain [132]. The second cleft separating the motor domains and the third domain is predominantly negatively charged and can therefore hold the positively charged single DNA strand [140].

To construct the elastic network of HCV helicase we have used the ligand-free apo

CHAPTER 3. CONFORMATIONAL DYNAMICS OF HEPATITIS C VIRUS HELICASE

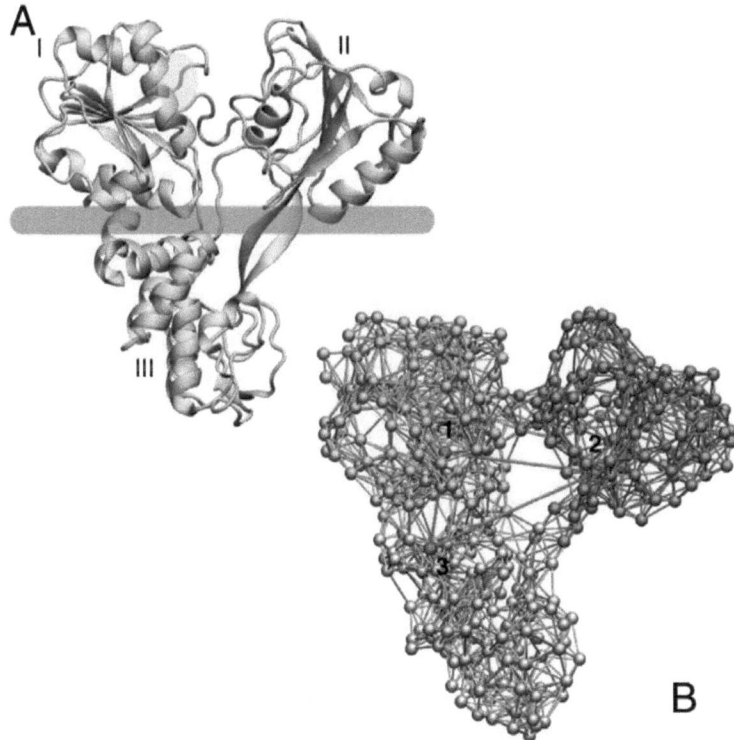

Figure 3.1: **HCV helicase and its elastic network.** The ribbon representation (A) is for the apo form of the protein. Three domains I, II, and III are indicated. The approximate location of the ATP binding pocket is shown by yellow color. The green stripe highlights the position of the DNA binding cleft. The elastic network (B) consists of 443 particles connected by 2235 links. Domains I, II, and III are colored as orange, blue and silver. Three particles 1, 2, and 3, corresponding to residues Thr269, Thr411, and Trp501, are chosen as labels. Distances between them (i.e. the lengths of the green lines) are used for visualization of conformational motions.

structure of the protein [133] deposited in the Protein Data Bank (PDB ID code 1HEI). This entry provides two independent data sets. We have chosen chain A since it yields more complete structural information, containing N=443 amino acid residues. Using this data, the elastic network has been constructed by treating each amino acid (the entire residue) as a single site represented by a point-like particle. Each particle is placed at the equilibrium position of the α-carbon atom in the main chain of the respective residue, denoted by $\vec{R}_i^{(0)}$ for particle i. Then, two particles are connected by a deformable elastic spring if the distance between them is less than a prescribed interaction radius l_{int}. The connection pattern of the network is stored in matrix **A** with the elements $A_{ij} = 1$, if $d_{ij}^{(0)} = |\vec{R}_i^{(0)} - \vec{R}_j^{(0)}| < l_{int}$, and $A_{ij} = 0$, otherwise. As the interaction radius, we have chosen 8Å. This value is large enough to prevent weakly connected residue groups in the network or even disconnected parts, but also not too large so that the network would be unrealistically stiff and large-scale motions could not be studied. In a deformed state, the positions of the particles $\vec{R}_i$ are changed and the lengths $d_{ij} = |\vec{R}_i - \vec{R}_j|$ of elastic springs are generally different from their natural length $d_{ij}^{(0)}$. All springs have the same stiffness constant κ. The elastic energy of the network is

$$U = (\kappa/2) \sum_{i<j} (d_{ij} - d_{ij}^{(0)}). \tag{3.1}$$

By construction, the energy minimum $U = 0$ corresponds to the experimentally known native conformation.

3.2 Probing of mechanical properties

Now that we have constructed the elastic network of the HCV helicase protein, its mechanical properties can be examined. We will probe the dynamical properties of the protein by exposing its elastic network to various initial deformations of its shape. The internal forces will then guide this elastic object to its original shape. By analyzing various relaxation processes, which start from different initial deformations, the properties of relaxation dynamics in the protein can be investigated.

Two different kinds of investigations will be performed. In the first study, relaxation starting from the deformations that have been obtained by perturbing the network structure globally will be considered. In the second study, our attention is focused on the perturbations localized in a particular region of the protein.

In the latter study, we wanted to take into account that in the actual operation cycles of ATP-driven protein motors, functional conformational motions are induced by the perturbations localized near the nucleotide binding site and involve binding of ATP, the hydrolysis reaction and the release of chemical products. The approximate nature of the elastic network description does not allow us to include specific chemical details of all such processes. Nonetheless, we can still probe the related conformational responses by considering perturbations localized at the residues which belong to the experimentally known conserved motifs and are expected to be involved in the ATP binding and hydrolysis reaction. The details of actual interactions between the HCV helicase and the ATP molecules are not known, but, assuming that local forces are generated in the ATP binding site, we can roughly imitate such interactions by the application of random forces in this local region.

We needed to generate large ensembles of initial network deformations. These deformations were obtained by applying random static forces to the beads of the network. Specifically, we have generated forces $\vec{f}_i$ acting on bead i, by choosing the components f_{x_i}, f_{y_i} and f_{z_i} as random numbers from the interval between -1.0 and 1.0 and, after that, rescaling the forces in such a way that the normalization condition $(\sum_i |f_i|^2)^{1/2} = C$ was always satisfied. In the first case of global perturbations, random static forces were applied globally distributed over all network beads, while, in the case of localized perturbations, forces were applied only to those network beads that corresponded to residues of the conserved ATP binding motifs. Those were the residue sequences $Gly^{207} - Ser - Gly - Lys - Ser - Thr$ (the Walker A motif) and $Asp^{290} - Glu - Cys - His$ (the Walker B motif).

To obtain the coordinates of beads in the initial deformed state of the network, we numerically integrated the equations of motion (2.9) in the presence of the forces for a fixed time t_f. This procedure was repeated to prepare a set of 100 initial network deformations, each arising from a different random configuration of forces. For each prepared initial deformation, we have checked that the springs were not excessively stretched, i.e. that plastic deformations were excluded. Namely, we have required that elongations of the springs did not exceed a threshold $1.5 \cdot d_{int}$ in the initial deformed states. In the simulations, numerical values $C = 1$ and $t_f = 10,000$ were used.

In the next step, we have switched off the applied forces and analyzed relaxation processes starting from different initial deformations, which have been thus prepared.

3.3 Visualization of conformational motions

The response to initial deformations of the elastic network consists in conformational relaxation motions induced by internal forces due to the deformed springs. To study such motions we needed to visualize the conformational dynamics. Although the dynamics involves motions of all protein residues, we have selected only three of them as labels to conveniently track the relaxation motions in the three-dimensional space. Conformational changes have been monitored by following the time evolution of the distances between the chosen labels. Thus, relaxation processes, each starting from a different initial deformation, were represented by trajectories in a three-dimensional space.

The labels have been chosen to belong to three different domains. Labels 1 and 2 lie in the motor domains I and II, whereas label 3 belongs to domain III. They correspond to the protein residues Thr^{269}, Thr^{411}, and Trp^{501}. For presentation of the pattern of relaxation trajectories, instead of the absolute distances between the labels, we have chosen normalized relative distance changes u_{12}, u_{23} and u_{13} between them, i.e, for example, $u_{12} = (d_{12} - d_{12}^{(0)})/d_{12}^{(0)}$.

For such variables, the origin of coordinates always corresponds to the equilibrium conformation of the network.

Figure 3.2 shows the relaxation pattern for the elastic network of HCV helicase. Grey trajectories start from different initial deformations obtained by applying global deformations, whereas relaxation processes starting from the states induced by the application of local forces in the ATP binding region are shown as red trajectories.

The first remarkable aspect to notice is that although the forces have been generated at random, the initial deformations obtained upon their application are actually far from being completely random.

We have found that it is almost impossible to induce relative motions of motor domain I with respect to domain III. On the other side, by the application of the same forces, large-scale relative motions of the motor domain II with respect to the rest of the protein can be easily induced. Relaxation trajectories could therefore be presented in a two-dimensional plot where the coordinates are normalized relative distance changes

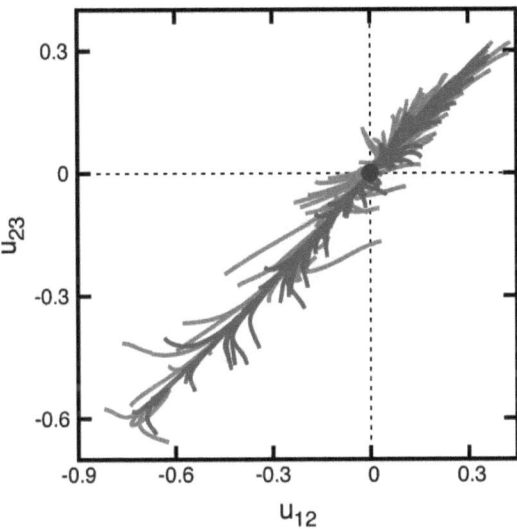

Figure 3.2: **Conformational relaxation in HCV helicase.** Gray lines show 100 relaxation trajectories in the plane of normalized distance changes u_{12} and u_{23} between the chosen labels, each starting from a different initial conformation obtained by the application of random static forces to all particles. Red lines display 100 relaxation trajectories, but starting from the initial conformations obtained by the application of static forces only to the particles in the ATP binding region.

between the two motor domains, represented by u_{12}, and between motor domain II and third domain, represented by u_{23}, respectively. As indicated by the starting points of the trajectories in Fig. 3.2, changes of up to 60% between the motor domain II and the rest of the protein can be induced by application of the forces.

Looking now at the relaxation trajectories for the ensemble of different initially deformed states, we see that, even though most of the initial conformations corresponded to highly deformed states, the elastic network of the HCV helicase protein could always return to its equilibrium shape. No meta-stable states could thus be found. Starting from the various deformed states, after short transients, all relaxation trajectories converged to an attractive path, that, once reached, guided the deformed network back to the original equilibrium conformation.

The dynamics along such attractive bundle is low-dimensional and corresponds to the ordered motions of the motor domain II with respect to the other two protein domains. Along the relaxation paths, the distances between motor domain II and the two other domains change persistently, so that this motion be characterized as of hinge type.

When perturbations localized in the ATP binding region were considered, we have found that, although being spatially confined, they could induce large initial deformations of the network as well, and, the resulting relaxation trajectories were quite similar.

We conclude therefore that, in response to different initial perturbations, the HCV helicase protein tends to produce the same well-defined kinds of conformational motions. These motions are generic, they did not sensitively depend on a particular initial deformation. The motions approximately represent well-coordinated rotations of the mobile motor domain II with respect to the motor domain I and the domain III, which together behave like a single rigid object.

By probing the dynamical relaxation properties of the HCV helicase elastic network, we have thus demonstrated that this motor protein is able to perform robust and well-coordinated internal motions of its domains. In earlier investigations, it has been shown that various other motor proteins, such as F1-ATPase, myosins II and V and kinesin KIF1A, share the same property [118, 119]. They can perform ordered conformational motions that are robust against perturbations and exhibit only weak dependence on the initial conditions. This property is fundamental; it underlies the ability of molecular machines to reproduce the same coordinated internal movements in each next operation cycle. Such a dynamical behavior might have been evolved as a result of biological evolution which, under the evolutionary pressure, has favored special dynamical properties of the motor proteins.

3.4 Modeling of ATP-dependent operation cycles

After probing relaxation dynamics of the HCV helicase protein by exposing its elastic network to various mechanical perturbations and subsequently tracking its response, we go a step further and extend the elastic network model in such a way that it allows us to reproduce the ATP-induced conformational changes underlying the operation of the HCV helicase motor. To do this, we first needed to find a way how to model the binding of an ATP molecule to the equilibrium conformation of the HCV helicase protein.

The ATP is a relatively complex molecule which consists of the adenine part, connected to a ribose sugar moiety, and the tri-phosphate tail. Binding processes of ATP to the protein rely on complicated interactions with chemical details that cannot be resolved by us. It would be, however, natural to assume that, when an ATP molecule arrives, its interactions with the protein lead to the appearance of some local deformations in the ATP binding pocket region. In the coarse-grained elastic-network model with residues replaced by single beads, the chemical structure of ATP and details of its interactions with the residues cannot be resolved. Our results described in the previous section suggest, however, that such details may actually not be important when ligand-induced conformational motions are considered.

To remain at the coarse-grained description level on which our protein modeling is based, the binding of an ATP molecule was also taken into account in an approximate way. The binding process was imitated by placing an additional particle (the substrate ligand) into the ATP binding pocket of motor domain I. The new particle established four elastic links to the neighboring network nodes inside this pocket (see Fig. 3.3). It had the same mobility as all particles in the protein network and the new links were of the same kind as the network links, i.e. they were assumed to have the same stiffness. The newly established links were initially deformed (stretched), roughly emulating the transfer of energy upon the binding event.

The network-substrate complex has the elastic energy

$$U_{complex} = (\kappa/2) \sum_{i<j} A_{ij}(d_{ij} - d_{ij}^{(0)})^2 + (\kappa/2) \sum_{i=c_1,c_2,c_3,c_4} (d_{i,N+1} - l_{nat})^2. \quad (3.2)$$

The first term is the energy of the free network and the second term is the energy of ligand-network interactions. Here, $c_1, \ldots, c_4$ are the indices of the particles to which the ligand, having the index $N+1$, is bound and l_{nat} are the natural lengths of the new elastic links.

Initially localized in the four new links, the deformations spread over the network and induce motions of the motor domains. These motions represent conformational relaxation of the network-substrate complex towards its equilibrium conformation. They end when this stationary state is reached.

3.4. MODELING OF ATP-DEPENDENT OPERATION CYCLES

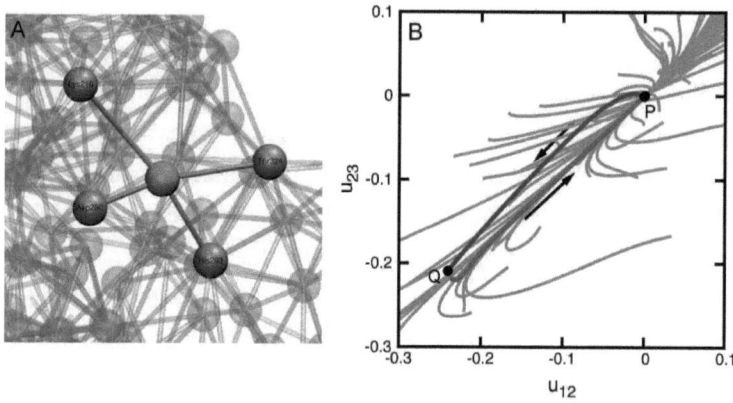

Figure 3.3: **Modeling of ATP binding.** (A) Binding configuration of the substrate ligand (green ball) to the ATP binding pocket of motor domain I. The ligand establishes four links to residues Lys^{210}, Asp^{290}, His^{293} and Thr^{324}, all belonging to the conserved residues of the ATP binding pocket. (B) Conformational motion $P \to Q$ is induced by placing the substrate ligand particle into the ATP binding pocket; the reverse motion $Q \to P$ is observed when the ligand, after its conversion from substrate into product, is released. For comparison, gray lines (the same as in Fig. 3.2) show the relaxation trajectories.

Figure 3.3 displays the relaxation trajectory $P \to Q$ of the conformational motion induced by binding of the substrate ligand. The most notable structural change that we observe is a large-amplitude relative motion of the two motor domains with respect to each other. Upon binding of the substrate, they move towards each other, so that the spatial gap between them gets closed. In the final conformation, corresponding to the equilibrium of the network-substrate complex, the distance between the labels of the two motor domains is shortened by approximately 25% as compared to the equilibrium distance between them.

It has turned out that the amplitude of the induced motion, i.e. the intensity of response of the HCV helicase protein to binding of the substrate ligand, sensitively depends on the choice of the four residues to which the links are established. We have found that binding of the ligand to residues Lys^{210} from the Walker A motif, Asp^{290} and His^{293} from the Walker B motif, and especially to Thr^{324} from the conserved $Thr^{322} - Ala - Thr$ sequence results in a pronounced, large-amplitude domain motion. The first three of these residues have been previously identified to play a key role in

ATP binding and hydrolysis based on an electrostatic analysis of this protein [140], and the residues of the $Thr^{322} - Ala - Thr$ sequence are also likely involved in the ATP binding [133]. The ligand binding configuration is shown in Fig. 3.3. Initially, the links established by the substrate ligand have lengths $d_{i,N+1} = (4.64, 6.48, 7.71, 5.87)$ (Å). Their natural lengths are all chosen to be the same and equal to $l_{nat} = 1$ Å, so that these additional links are initially stretched. By the time $t = 20,000$, the stationary state of the network-substrate complex, corresponding to the closed conformation, is reached.

The conformational relaxation, induced by substrate binding should have essential consequences. On the left side of motor domain II, an essential conserved residue group is present. Only when this atomic group comes into contact with the ATP, bound on the right side of motor domain I, the hydrolysis reaction converting ATP into ADP and a phosphate becomes possible [134].

The observed substrate-induced motion of domain II towards domain I brings into contact these two motor domains, so that the contact between the above-mentioned atomic group and the ATP can indeed be established. The hydrolysis reaction should take place at this stage, followed by the release of chemical products. Again, these chemical processes are beyond the descriptive power of the coarse-grained elastic network model. However, we can still roughly imitate them in our model.

To do this, we assume that, as the equilibrium state of the network-substrate complex is approached, the nature of the ligand becomes changed. It gets converted from the substrate into the product particle. In contrast to the substrate ligand, we assume that the product ligand has no interactions with the protein. Hydrolysis and product release are thus implemented in the model by just cutting the links that connect the ligand to the network and removing the product particle, once the equilibrium state of the complex has been reached.

After the release of the product particle, the free network finds itself in a conformation different from its true equilibrium state. Starting from this initial state, relaxation of the free network back to its equilibrium conformation begins. As we have seen when probing the dynamical properties of the HCV helicase in the previous section, such relaxation process, leading back to the original equilibrium state, will involve coordinated

3.4. MODELING OF ATP-DEPENDENT OPERATION CYCLES

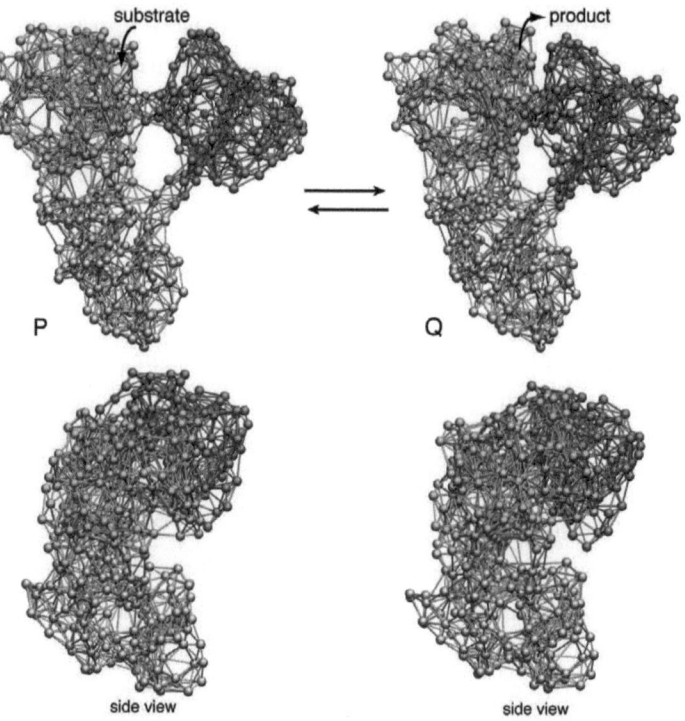

Figure 3.4: **Cyclic ligand-induced conformational motions in HCV helicase.** Upon binding of the substrate ligand to the equilibrium state of the network, the HCV helicase protein undergoes a conformational change from the open-shape conformation P, with the two motor domains being spatially separated, to the closed-shape conformation Q with the motor domains being adjacent. In this conformation, corresponding to the equilibrium state of the network-substrate complex, the substrate is converted into the product particle. Its subsequent removal induces relaxation of the free network accompanied by domain motions from the closed conformation Q back to the open conformation P.

motions of protein domains.

Under this back relaxation process, corresponding to the trajectory $Q \to P$ in Fig. 3.3, the motor domain II moved back away from motor domain I, restoring the initial configuration. By the time moment $t = 50,000$, one substrate-induced operation cycle of the HCV helicase protein becomes completed in our simulations and the next cycle, initiated by binding of another substrate ligand, can then take place.

Snapshots from a simulation showing the ligand-induced conformational motions in HCV helicase are provided in Fig. 3.4.

CHAPTER 3. CONFORMATIONAL DYNAMICS OF HEPATITIS C VIRUS HELICASE

Thus, through an extension of the elastic network model, incorporating interactions that mimic binding of an ATP molecule, hydrolysis and product release, we are able to reproduce the ATP-dependent cycle of the HCV helicase protein. The relaxation processes induced by substrate binding and product release, correspond to the conformational motions of the two motor domains which consist of a switching from the open to the closed conformation when the substrate is present and vice versa after the product became released.

The opening and closing motions of the two motor domains are one important aspect of the simulated cyclic relaxation dynamics. A more detailed examination of the induced relaxation motions actually reveals that they have another important component. This other aspect becomes apparent only when the ligand-induced cyclic motions are viewed in the side perspective the protein, focusing on the structural changes in the DNA binding cleft region (see Fig. 3.4).

In the side view, one can see that the two motor domains have roughly the shapes of claws confining the DNA cleft at two opposite ends. When ligand-induced conformational changes are considered, in can be noticed that, after binding of the substrate ligand, the motor domain I gets lifted and thereby widens the DNA cleft on one side, while at the same time the motor domain II quickly moves down and thus is narrowing the DNA cleft on its other side. After the conversion of the substrate ligand into the product and its subsequent release, opposite domain motions are observed.

These observations suggest that a hands-on hands-off mechanism of grip control of the HCV helicase on the DNA strand is possible. As a result of substrate binding to the protein, the motor domain II is able to rapidly establish a tight grip on the DNA (hand on), whereas the motor domain I is weakening the grip (hand off). It is the other way round after the hydrolysis and product release.

Cyclic large-amplitude concerted motions of the two motor domains together with a possible grip control on the DNA strand already suggest a certain mechanism used by the protein to translocate along the single DNA strand. However, further investigations were needed to clarify the way of locomotion along the nucleic acid strands. To this end, the model has been further extended. Besides modeling of interactions with the ATP molecules, the extended model had to include the dynamics of DNA strands and their interactions with the HCV helicase. The formulation and the analysis of the extended model are the subject of the next chapter.

Chapter 4

Entire operation cycles of Hepatitis C virus helicase

The use of the elastic network model to investigate conformational relaxation motions in the HCV helicase has already led us to some insights into the dynamical properties of this important protein motor. We have learned that coordinated domain motions are underlying the organized dynamics of this protein. They are used to execute robust operation cycles that are related to ATP binding and hydrolysis and consist of large-amplitude closing and opening motions of the cleft between the two motor domains together with more refined motions that control the dynamics of the DNA binding cleft.

To find out how such motions can lead to steady translocation of the helicase along the DNA strand, our investigation will be further extended to include also a dynamical description for the DNA molecule and interactions between the helicase and DNA.

4.1 Modeling of DNA

The DNA, the carrier of genetic information, is a highly complex molecule consisting in its commonly present duplex form of two single strands that are paired by interactions between complementary bases adenine/thymine and guanine/cytosine, respectively. These two strands are typically wound around each other giving rise to a helical arrangement of the duplex DNA in its stable form, the so called B-Form.

While detailed descriptions of DNA and its dynamics are possible, using them would be inconsistent with the otherwise coarse-grained nature of descriptions em-

4.1. MODELING OF DNA

ployed in this study. For our purposes, it is more important to concentrate on a proper description of the mechanical properties of DNA, rather than on taking into account its detailed chemical interactions. To remain at the approximate level on which our dynamical description of the protein relies, we have modeled the DNA – its structure and interactions – also within a coarse framework.

A single DNA strand is modeled by us as a semi-flexible polymer chain consisting of identical beads, each replacing a phosphate-sugar-base unit which represents the principal building blocks of nucleic acids. A bead is connected to its neighboring beads by flexible links. Figure 4.1 shows the setup of a single DNA polymer chain.

The chain of beads representing one single DNA strand is allowed to undergo deformations such as stretching and bending. Its total elastic energy is

$$U_{sDNA} = (k_{el}/2) \sum_{i=1}^{M-1} (|\vec{r}_i - \vec{r}_{i+1}|^2 - \sigma) + k_{bend} \sum_{i=1}^{M-1} (1 - \cos\theta_i)^2, \quad (4.1)$$

where M is the number of beads in the polymer chain. The first term in this equation, a harmonic bonding potential, takes into account elastic deformations of links in the chain. Here, $\vec{r}_i$ is the position vector of bead i, σ is the native length of the links, and k_{el} is the stiffness constant, representing the stiff sugar-phosphate backbone of DNA. The second term is the elastic bending energy of the chain, with θ_i being the angle between two adjacent links in the chain and k_{bend} being the bending stiffness constant. The angle is defined as

$$\cos\theta_i = \frac{(\vec{r}_i - \vec{r}_{i-1}) \cdot (\vec{r}_{i+1} - \vec{r}_i)}{|\vec{r}_i - \vec{r}_{i-1}||\vec{r}_{i+1} - \vec{r}_i|}. \quad (4.2)$$

In the same way as done for the protein dynamics, the relaxational motion of the beads in the DNA chain is described by a set of differential equations

$$\frac{d\vec{r}_i}{dt} = -\gamma \frac{\partial U_{sDNA}}{\partial \vec{r}_i}, \quad (4.3)$$

which are numerically integrated to obtain the time evolution of the bead positions in the chain. For simplicity, we have assumed that beads of the DNA had the same

CHAPTER 4. ENTIRE OPERATION CYCLES OF HEPATITIS C VIRUS HELICASE

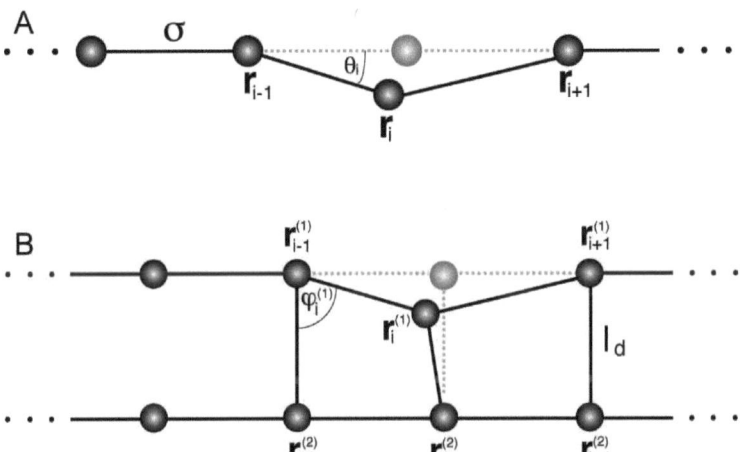

Figure 4.1: **Modeling of the DNA molecule.** (A) Semi-flexible chain model of a single DNA strand. (B) Setup of the duplex DNA composed of two single chains connected by breakable bridge-links.

mobility as the particles that form the protein network, i.e. $\gamma = \Gamma$.

Several aspects of the mechanical properties of DNA are known from the force-probing experiments (see e.g. [77]). The sugar-phosphate backbone of a single DNA strand, for instance, is relatively stiff and cannot be stretched by large amounts, meaning that the distance between two beads in the semi-flexible chain should not be allowed to change much. The DNA strand is also rigid with respect to the bending behavior with a typical persistence length of about 10 base-pairs. Although no fine-tuning of the model parameters has been undertaken by us, we have still taken those known aspects into account by making the DNA chain much stiffer than the elastic network of the helicase protein. In our simulations, we have used numerical values $k_{el}/\kappa = 20$ for the relative stiffness constant and $k_{bend}/\kappa = 15$ Å^2 for the relative bending stiffness constant. For the beads separation in the chain we have chosen $\sigma = 6.5$Å.

To further demonstrate how the HCV helicase is able to perform the separation of the duplex DNA into its single-strand components, we have modeled the duplex part as a setup of two single polymer chains that are connected by additional links that bridge opposite chain beads (see Fig. 4.1). These links shall mimic the attractive interactions between complementary bases of the two chains that hold together the duplex form.

On the other hand, also their breaking should be considered when the strands become spatially separated. In our approximate description, only effective interactions could be accounted for and the dynamics of bridge-links may be chosen to be described by a fairly arbitrary potential that has a well and becomes flat for large-enough separations between the beads (i.e. the interactions drop to zero and the links disappear). In computer simulations, the Lennard-Jones potential is often used for an effective description of DNA bridge-links. We have however decided to employ the Morse-potential in our simulations.

The full energy of this duplex DNA system is the sum of the energies of the two single strands and of the interaction potential, i.e. $U_{dDNA} = U_{sDNA}^{(1)} + U_{sDNA}^{(2)} + U_{int}$. Here, superscripts 1 and 2 are used to distinguish between the two chains. The interaction potential is

$$U_{int} = D \sum_{i=1}^{M} [\exp(-a\Delta_i) - 1]^2 + \kappa_s \sum_{i=1}^{M-1} [(\cos \varphi_i^{(1)})^2 + (\cos \varphi_i^{(2)})^2]. \quad (4.4)$$

The first term is a sum of the Morse potentials, where $\Delta_i = |\vec{r}_i^{(1)} - \vec{r}_i^{(2)}| - l_d$ is the elongation of a bridge-link i. The second term depends on the relative orientation of of the bridge-links and prevents shear motion of the two DNA strands with respect to each other, taking into account the shear stiffness of the duplex DNA. Here, the angles are defined as

$$\cos \varphi_i^{(1)} = \frac{(\vec{r}_{i+1}^{(1)} - \vec{r}_i^{(1)}) \cdot (\vec{r}_i^{(2)} - \vec{r}_i^{(1)})}{|\vec{r}_{i+1}^{(1)} - \vec{r}_i^{(1)}||\vec{r}_i^{(2)} - \vec{r}_i^{(1)}|}. \quad (4.5)$$

The Morse potential becomes flat if the elongation of a bridge-link is much larger than the characteristic length $1/a$. Thus, the force is vanishing and the link becomes effectively broken which allows us to model the DNA unzipping process later.

Some parameters have been fixed using the known geometry of the duplex DNA while others have been simply roughly estimated. In the simulations, numerical values $a = 4.5$ Å^{-1}, $D/\kappa = 6.67$ Å^2, $l_d = 20$ Å, and $\kappa_s/\kappa = 0.05$ Å^2 were used. To simplify calculations, we have introduced a cutoff at $\Delta_i = 5.5$ Å, assuming that bridge interactions disappear completely for larger separations between two opposite beads.

The position of each bead in the duplex DNA and its temporal changes are again obtained by numerically integrating the set of differential equations,

$$d\vec{r}_i/dt = -\gamma \partial U_{dDNA}/\partial \vec{r}_i. \tag{4.6}$$

As it has already been said above, our aim in this thesis is not to perform a detailed study of the dynamical properties of the DNA. Investigations of this important molecule actually present a research topic on its own. The modeling undertaken by us should be rather viewed in the context of the operation of the hepatitis C virus helicase on the duplex DNA, which is the translocation along it and its separation. Therefore, we have employed the coarse-grained description of nucleic acids, which is similar to the approaches used by other authors [141, 142, 143]. We have also applied further simplifications such as neglecting the helicity of the duplex DNA or the use of an auxiliary anti-shearing potential. There are however also some justifications for this. A certain class of proteins, the topo-isomerases, is able to remove the torque of DNA before the helicases can unzip them so that we can indeed assume that near the unzipping region the duplex DNA is almost flat. On the other hand, the unzipping of duplex DNA merely by shearing, although it has been modeled also, is of less biological importance [144].

4.2 Translocation mechanism of HCV helicase

In the previous chapter we have shown that binding of s substrate particle to the HCV helicase elastic network (mimicking ATP binding) and the subsequent release after its conversion into the product particle (mimicking ATP hydrolysis and the product ejection) first induce a motion from the open to the closed conformation of the protein, making hydrolysis possible, and then, upon the release of the product ligand, reset the original open protein state. These large-scale conformational changes are also accompanied by controlled motor domain motions that affect the DNA binding cleft, suggesting a hands-on hands-off mechanism of the grip control. Now, we go further and try to understand how the cyclic changes in the HCV helicase can be used by the protein to walk along a single strand of DNA.

It is clear that the interactions between DNA and the protein are complicated and to determine them accurately would require a modeling that is beyond our scope. In our simplified model, we can only phenomenologically imitate such interactions. To do this, we have assumed that elastic links between certain particles of the two motor domains, located near the DNA binding cleft, and the beads forming the DNA polymer chain could be established.

To consider interaction between the helicase and the DNA, we need to place the single DNA chain into the DNA binding cleft. From the crystallographic studies of the HCV helicase it is known that in the DNA-bound state the two motor domains are approximately separated by the distance of three nucleotides [134]. We have placed the single chain of DNA roughly in the middle of the binding cleft and adjusted the distance between two consecutive beads in the DNA strand to fit this requirement ($\sigma = 6.5$ Å, see Fig. 4.1).

Moreover, it is known that major interactions between the helicase and the bound single DNA strand are not distributed over the whole binding cleft, but rather confined to only a number of specific conserved residues [134]. Such key residues that are known to establish major contacts to the sugar-phosphate backbone of the DNA are Thr^{269} from motor domain I and Thr^{411} from motor domain II [134]. We have taken this into account in our modeling and restricted possible interactions between the helicase and DNA to the particles that corresponded to those two residues. The interactions were mediated in our description by additional elastic links. They were established or disappeared depending on the presence of the ligand.

When the substrate-ligand binds to the helicase network, i.e. when the motor domain II moves down narrowing the DNA binding channel and thus establishes a tight grip on the DNA, the distance between Thr^{411} and the DNA chain decreases. To model the tight gripping, we have therefore assumed that a relatively stiff link between the residue Thr^{411} and the nearest bead in the DNA chain becomes established after binding of the substrate-ligand.

When the product-ligand is released from the network, i.e. when motor domain I is about to establish a tight grip on the DNA while the motor domain II loosens its grip, the link between the DNA chain and residue Thr^{411} from domain II is removed and a link between the residue Thr^{269} from domain I and the DNA chain is created in the

CHAPTER 4. ENTIRE OPERATION CYCLES OF HEPATITIS C VIRUS HELICASE

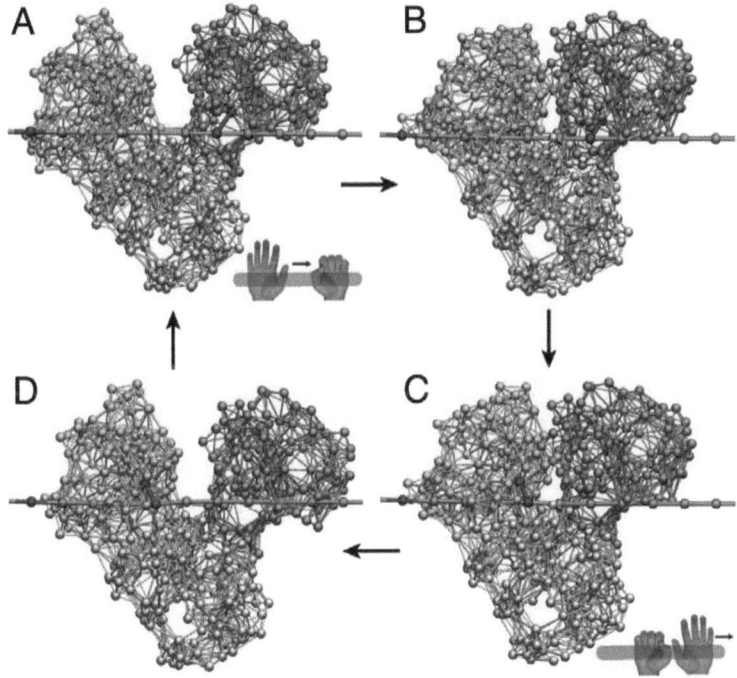

Figure 4.2: **HCV helicase Translocation mechanism.** Four consequent snapshots (A-D) within a single translocation cycle are shown. After each cycle, the protein conformation is repeated, but the protein becomes translocated by one nucleotide along the DNA strand. The bottom view is chosen here, with the motor domains I and II being above domain III. Red links show connections between the protein and the DNA strand. One bead in the DNA chain is colored red to better illustrate the translocation.

same manner as described above.

The interaction mediated through the link depends on the length of the link and its relative orientation. The length dependence is described by a harmonic potential with the relative stiffness constant $k_{h,DNA}/\kappa = 5$ and the natural length $l_{h,DNA}^{(0)} = 5$ Å. Additionally, the interaction depends on the relative orientation of the link with respect to the DNA chain. The orientation interaction is stiff, and, as we have assumed for simplicity in our simulations, a definite angle between the link and the DNA chain is always maintained. Furthermore, the first and the last of the DNA chain were always immobilized in the simulations.

4.2. TRANSLOCATION MECHANISM OF HCV HELICASE

With such model extensions, we could successfully reproduce the mechanism by which the HCV helicase is able to move along the DNA strand in a structurally resolved way (see Fig. 4.2). In the equilibrium ligand-free state, the DNA is attached by a link to the left motor domain I. Binding of the substrate-ligand induces a quick lift of the claw formed by domain I (hand-off) so the link attaching DNA to domain I disappears. At the same time, the claw of domain II moves down (hand-on) and a link attaching this domain to DNA becomes formed. After that, the slower relative motion of motor domain I towards motor domain II which is holding the DNA strand, is taking place. When the equilibrium of the ligand-network complex is reached, the product-ligand is removed. Now, the grip is changed. The left motor domain I moves down to the DNA and a link attaching it to one of the chain beads is established, whereas the right motor domain II moves up and the link between it and the DNA strand disappears. Then, motor domain II slowly returns to its position by moving away from motor domain I, but without carrying the DNA strand with it.

Thus, after each ligand-binding and ligand-release cycle, the protein returns to its initial conformation, but it becomes translocated by one nucleotide along the DNA strand (see Fig. 4.2).

The style of locomotion of the HCV helicase protein along the DNA is reminiscent of the mechanism by which worms deform themselves to crawl along the stalks of the plants. As our findings show, the helicase also crawls along the DNA strands. The ATP-dependent deformations in the helicase consist in motions that alternately bring together and separate two opposite modules (the motor domains). Their integrated gripping (ratcheting) mechanism accounts for the conversion of these internal motions into directed base-by base locomotion along the DNA strand. Because of such analogy, the described mechanism of translocation is often referred to as the *ratcheting inchworm translocation* [60, 145]. Previously, this mechanism has been proposed based on the experimental data [134, 136]. It could not, however, been theoretically confirmed so far.

CHAPTER 4. ENTIRE OPERATION CYCLES OF HEPATITIS C VIRUS HELICASE

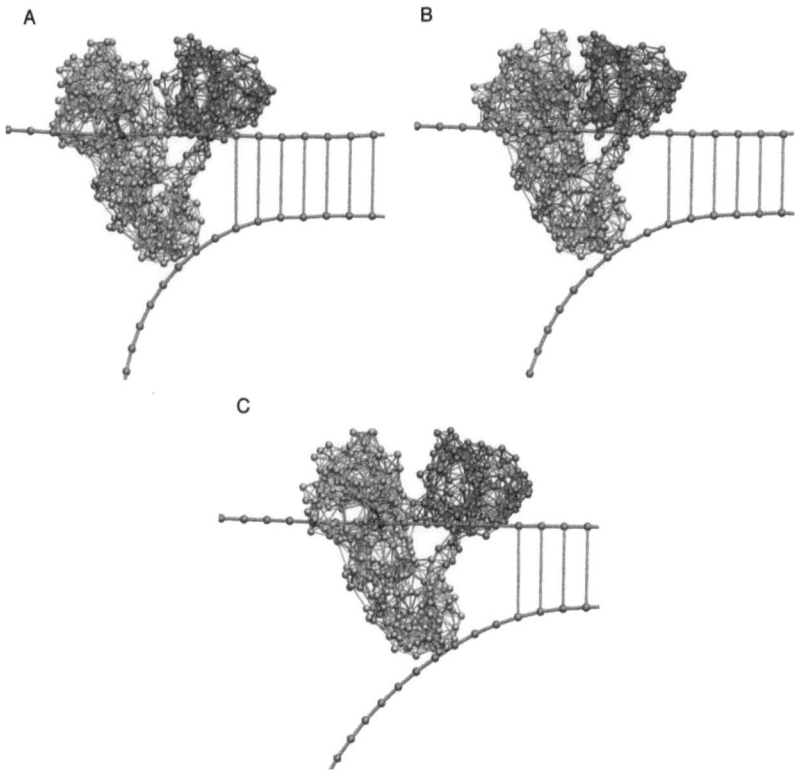

Figure 4.3: **Duplex DNA unzipping by HCV helicase.** Three consequent snapshots (A-C) from a simulation showing ratcheting inchworm translocation of HCV helicase and mechanical unzipping of duplex DNA.

4.3 Duplex DNA unzipping

The two motor domains in HCV helicase serve as basic modules that make ratcheting inchworm locomotion along the DNA with a step-size of one base per ATP possible. The role of the third protein domain, that, as revealed by the dynamical probing, is rigidly connected to motor domain I, is however still unclear. During the translocation of the motor modules this domain will be always dragged with it by the motor domain I. It is therefore plausible that this domain will play a crucial role the the separation process of duplex DNA.

We have further extended our model to describe also interactions with the duplex DNA with the aim to demonstrate duplex DNA unzipping by the HCV helicase. The duplex DNA has been modeled as two single chains bridged by breakable links as described in section 4.1 of this chapter. The duplex DNA is placed in such a way that the upper single chain fits into the designated binding cleft (as before, when translocation was considered) while the lower chain is oriented to lie below the two motor domains.

In the next step, we had to arrange the initial setup for the helicase and the duplex. To do this, we had to manually open the duplex at one side to prepare the fork at which the helicase can start its operation. This has been done by applying an appropriate force to the first bead in the lower DNA chain and mechanically unzipping the duplex for a while, so that the first couple of bridge-links get broken. In the prepared configuration, the free end of the lower DNA strand is positioned to the right of the third helicase domain which represents the most likely alignment as evidenced by experimental studies [132].

Little is known about the interactions between the HCV helicase and the duplex DNA because detailed experimental structures of such a complex are still lacking. Within our mechanical model and based on previously suggested possible mechanisms [136], we can however assume the following situation:

As the two motor domains are repeatedly translocating along the upper DNA chain, the third domain encounters the free end of the lower chain at the DNA fork and interactions between them should occur. It is obvious that the third domain should press against the lower strand; or, in other words, the third helicase domain is seen as a barrier for the lower DNA chain which cannot be penetrated. Alternatively, the consequence would have been partial unfolding of the third domain which is not known to take place [132]. Obviously, if some interactions between the third helicase domain and the lower DNA strand need to be introduced in a phenomenological way into the model, they must be of repulsive kind.

To account for the mechanical action of the third helicase domain, we have assumed that the lower DNA chain is subject to an overall repulsive potential emerging from all particles of the third domain. It reads as

$$U_{rep} = \tilde{D} \sum_{i,j} \{\exp[-a(|\vec{r}_i^{(2)} - \vec{R}_j| - r_{int})] - 1\}^2 \Theta(r_{int} - |\vec{r}_i^{(2)} - \vec{R}_j|). \quad (4.7)$$

Here, index $i = 1, \ldots, M$ is used for the beads of the lower DNA chain and j for the particles of domain III. The overall potential is a sum of repulsive Morse potentials with the interaction radius r_{int}, meaning that interactions between a bead of the lower DNA chain and a particle of domain III are present only if the distance between them is below r_{int}. This cutoff is implemented by the step function $\Theta(d) = 1$, if $d > 0$ and $\Theta(d) = 0$, else. Numerical values $r_{int} = 2$ Å, $\tilde{D}/\kappa = 25$ Å^2, and $a = 1$ Å^{-1} have been used.

In addition to short-ranged repulsive forces, electrostatic interactions between DNA and the protein can be important. Specifically, they may be responsible for keeping the lower DNA strand below the third helicase domain and preventing it from side motions (and eventually slipping away). Such interactions were not included in the model. To hold the free end of the lower DNA strand below the third helicase domain, a simple prescription was employed in the numerical simulations: After calculating all the forces acting on beads of the lower DNA chain, their components pointing out of the initially defined plane of the DNA duplex were deleted in each integration step. In other words, opposite forces mimicking a steep confining potential and thus preventing deviations from the chosen plane, were always virtually present. Additionally, the anti-shearing forces, corresponding to the second term in equ. (4.4), were taken into account only for base-pairs in the duplex part of the DNA.

With these model extensions, we were able in our simulation to watch the HCV helicase motor in action. We have followed three translocation cycles of the helicase in a long computer simulation and observed how progressive base-by-base inchworm translocation along the upper DNA strand leads to duplex strand separation. For the snapshots of the simulation, see Fig. 4.3.

As the two motor domains were translocating themselves over the upper strand of DNA, helicase domain III became dragged into the free space at the duplex fork. Due to the repulsive interactions, forces that try to prevent contact of domain III and the lower strand were produced which means that this strand was pressed away. This implies that also forces tearing apart the two DNA strands were generated. As they were built up, stress was accumulated in the links bridging the strands and eventually one bridge-link was getting broken, unzipping the DNA by one base-pair. Our simulations show that the function of helicase domain III is that of a wedge that is pressed between the two DNA strands and mechanically separates them while the motor domains translocate along the upper strand. We find that after three translocation steps the HCV helicase has managed to break three base-pairs, one in each cycle.

In the previous and the present chapters, we have presented a number of results of the HCV helicase modeling. Since they constitute a major contribution to this thesis, a preliminary summary and discussion is in order at this point. In this project, our aim was to understand the operation of a typical motor protein – the helicase of the hepatitis C virus – that plays a crucial role in the replication cycle of the virus. To do this, simplified descriptions for the protein, the DNA, and for the interactions between them and with the ATP ligand have been used in our study. The helicase was modeled as a network of identical particles connected by elastic links. DNA was also described by an elastic chain of identical beads. The binding of ATP, its hydrolysis, and the release of products were only roughly accounted for by assuming that a fictitious substrate ligand particle gets bound to the network and is converted into the product, which is then immediately ejected. Interactions between the helicase and DNA were phenomenologically incorporated into the model, assuming that elastic links connecting one of the network particles and a bead in the DNA chain become established or removed. Additionally, thermal fluctuations and interactions with the solvent were omitted in our description.

It is remarkable that even such a simple purely mechanical model could reproduce the principal operation of HCV helicase and allowed us to trace entire cycles of this molecular motor in a structurally resolved manner. Furthermore, our simulations indeed confirm that ratcheting inchworm translocation and spring-loaded DNA unwinding, previously proposed based on experimental data as the operation mechanism of HCV helicase [134, 136], can take place.

CHAPTER 4. ENTIRE OPERATION CYCLES OF HEPATITIS C VIRUS HELICASE

After our results had been published [139], two studies have reported several crystallographic snapshots of the HCV helicase complexed with non-hydrolyzable ATP-analogs and DNA, RNA, respectively [137, 138]. The obtained results agree well with our predictions. A brand new experimental work has also reported on single-base pair unwinding of the HCV helicase agreeing with our *in silico* predictions [146].

Chapter 5

Conformational motions in superfamily 2 helicases

In the introduction, the importance of helicase proteins and their classification have been outlined. In the subsequent chapters, we have shown how approximate models can be used to successfully trace entire operation cycles of the hepatitis C virus helicase in structurally resolved dynamical simulations. We have seen that the property of performing ordered conformational motions in this protein is fundamental since such motions allow this molecular motor to reproduce the same coordinated internal movements in each operation cycle.

The aim of a further investigation carried out by us was to check whether such ordered and well-defined conformational motions can be also found for other helicases in the superfamily 2. In contrast to HCV helicase, the considered proteins have been so far much less experimentally and computationally investigated and many essential aspects of their biological functions are still unknown. Therefore, we could not carry out the same detailed analysis for them as for the HCV helicase. Our studies were focused on probing mechanical properties of three other helicases and investigating their conformational motions by means of the methods similar to those used for the HCV helicase.

The results of this chapter have been published [147] and videos presenting these results can be accessed online via the PLoS ONE webpage.

5.1 *In silico* investigation of superfamily 2 helicases

As shown in the previous chapter, we were able to trace entire operation cycles of the HCV helicase in a structurally resolved manner by using an elastic network model for the protein and including its interactions with DNA and ATP. We have identified large-scale ordered motions inside the protein that bring together the two motor domains or spatially separate them depending on the presence of the ATP ligand. The switching between open and closed protein conformation can drive translocation and, as we have shown, HCV helicase can move along the nucleic acid chain by one base per cycle induced by the binding of one ATP molecule. It has been furthermore demonstrated that the third domain of HCV helicase acts as a wedge which is dragged between the two nucleic acid strands and mechanically separates them.

While significant experimental and theoretical progress in understanding the function of HCV helicase has been made, operation mechanism of other helicases in the same superfamily 2, especially those with limited structural data available, remain less clear. The open questions particularly refer to the ATP-dependent conformational motions inside these proteins and their coupling to the activity on the nucleic acid.

Whether the *one step translocation*, i.e. the base-by-base motion of the motor domains along the nucleic acid consuming one ATP in each cycle, represents the common mode of helicase locomotion is a topic of current debate [59, 145]. The mechanistic role of other structural domains in the processing of duplex DNA/RNA substrates by other helicases from superfamily 2 is also far from understood.

In the study presented in this chapter we have focused on the analysis of large-amplitude conformational motions in the superfamily 2 helicases. As in the previous case when functional motions in HCV helicase have been considered, investigations have been performed within the elastic network approximation. Moreover, similar methods have been used here to analyze mechanical responses of the selected proteins. This allows us to compare the properties of conformational motions for the chosen helicases with those of the HCV helicase.

Our investigations have been performed for the Hef helicase from *Pyrococcus furiosus* that manipulates fork-structured DNA forms [148], the Hel308 helicase from

CHAPTER 5. CONFORMATIONAL MOTIONS IN SUPERFAMILY 2 HELICASES

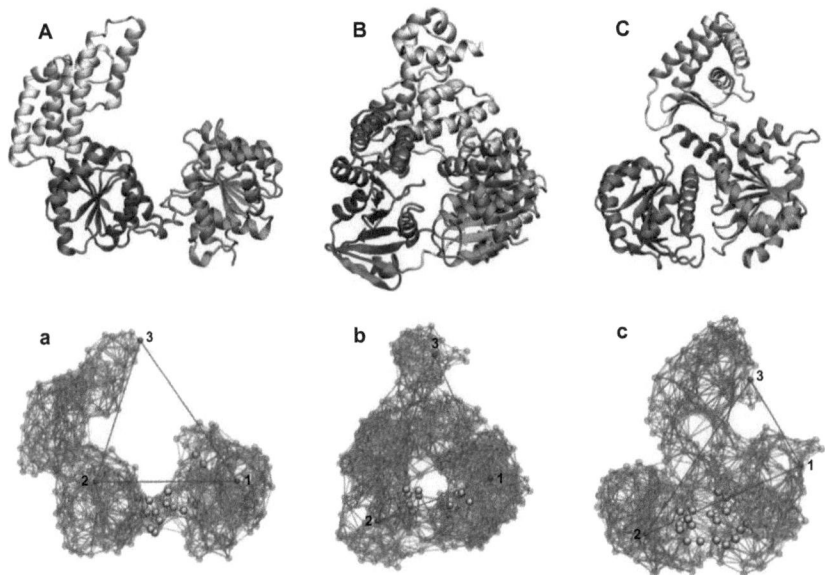

Figure 5.1: **Superfamily 2 helicases.** (A-C) Ribbon structures in the cartoon representation of Hef (A), Hel308 (B) and XPD helicase (C) and their respective elastic networks (a), (b) and (c). In the ribbon representation, the two motor domains are colored red (domain 1) and blue (domain 2). In the network representation, three labels 1, 2 and 3 are indicated. The distance between them, indicated by red lines, are used for visualization of conformational relaxation. The residues belonging to the conserved ATP binding motifs are highlighted as yellow spheres.

Archaeoglobus fulgidus involved in unwinding of lagging strands in replication forks and other branched nucleic acids [149], and the XPD helicase from *Sulfolobus tokodaii* that opens DNA duplex structures during transcription and repair processes [150]. We have constructed the elastic networks for the proteins and investigated their dynamical behaviour in response to various kinds of deformations.

Our primary aim was to find out whether the considered helicases, similar to other previously studied motor proteins [118, 119] and the HCV helicase, possess robust and well-defined conformational motions. Furthermore, we also wanted to discuss the possible functional role of such motions in the different helicases. Therefore, we have generated a large number of random initial deformations and considered conformational relaxation processes starting from them. In this way, robust conformational motions

5.1. *IN SILICO* INVESTIGATION OF SUPERFAMILY 2 HELICASES

could be identified for each of the proteins.

In the actual operation cycles of these molecular motors, functional conformational motions are induced by perturbations localized near the nucleotide binding site and involving binding of ATP, the hydrolysis reaction and the release of chemical products. The coarse-grained nature of the elastic-network models does not allow us to include specific chemical details of all such processes. Nonetheless, similar to the previous investigations of HCV helicase, we can still analyze the related conformational responses by applying various random perturbations (forces) to the residues which belong to the experimentally known conserved motifs and are expected to be involved in the binding of ATP and its hydrolysis reaction.

Figure 5.1 shows ribbon structures of the three considered helicase proteins and their respective elastic networks. The equilibrium conformations, taken from the Protein Data Bank, are 1WP9 (Hef, chain A), 2P6U (Hel308) and 2VL7 (XPD). Two motor domains and an additional third domain are present in all these proteins. It should be noted that these three helicases have not yet been co-crystalized with an ATP-analog and the corresponding ligand-induced structures are not known. Nonetheless, conserved residue motifs, common for all members of superfamily 2 helicases and associated with ATP binding have been identified [148, 149, 150]. In the studied helicases and the HCV helicase, they are located at the interface between the two motor domains (see Fig. 5.1), suggesting that ATP molecules bind in this region.

The same elastic network approximation as in the previous chapters will be used below. Amino acid residues are replaced by identical beads and effective interactions between them are taken into account through empirical harmonic potentials. Two beads are connected by a deformable link if the distance between them, extracted from the crystallographic data of the known equilibrium conformation, is within a prescribed interaction radius.

As in the previous study of the dynamics of HCV helicase, only slow conformational motions, with the time-scales of milliseconds or longer, will be considered here and, since then frictional dissipative forces dominate the conformational dynamics and moreover hydrodynamical interactions and thermal fluctuations are omitted by us, the motions inside the protein represent relaxation processes of its elastic network towards

CHAPTER 5. CONFORMATIONAL MOTIONS IN SUPERFAMILY 2 HELICASES

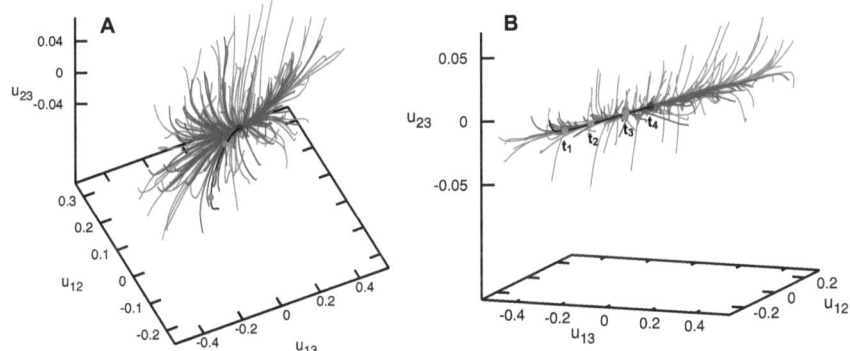

Figure 5.2: **Relaxation dynamics of Hef helicase.** Panels (A) and (B) show the relaxation pattern as a set of 100 trajectories in the space of normalized relative distance changes u_{12}, u_{23} and u_{13} between the three labels. Each trajectory starts from a different initial deformation that has been generated by applying random static forces globally distributed to all network beads (gray trajectories) or restricted to the beads of the ATP binding site (red trajectories). The viewpoint in (B) is chosen in such a way that the planar confinement of relaxation trajectories is best seen. Note that the scale for the distance changes u_{23} (the vertical axis) is much smaller than for the other axes.

the equilibrium configuration with the minimal elastic energy.

The dynamics of the three selected helicases has been probed by monitoring relaxation of their elastic networks after application of various initial deformations. The dynamics has been determined by numerically integrating the equations of motion, which yields the positions of network beads at all time moments. To visualize conformational changes we have selected three network beads as labels, each belonging to a different domain, and tracked the temporal evolution of the distances between them. In this manner, any conformational relaxation process could be characterized by a trajectory in the three-dimensional space of normalized distance changes between the labels with the origin corresponding to the equilibrium conformation.

Two different kinds of initial deformations were considered by us. The deformations of the first kind were obtained by perturbing the structure of the elastic network globally, i.e. by applying random forces globally distributed over all network beads. In the second case, perturbations were localized in the ATP binding pocket and initial deformations were generated by applying random forces only to those network beads that corresponded to the residues of the conserved ATP binding motifs. In both cases,

5.1. *IN SILICO* INVESTIGATION OF SUPERFAMILY 2 HELICASES

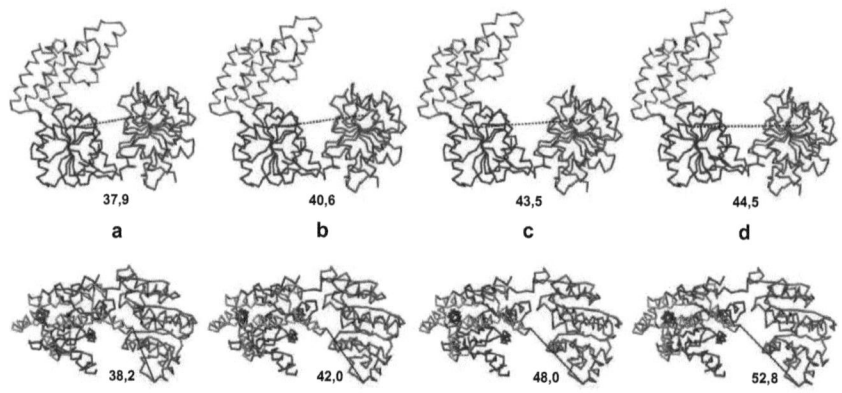

Figure 5.3: **Conformational motions in Hef helicase.** Consequent snapshots showing conformational changes in Hef helicase along the relaxation trajectory, highlighted in black in Fig. 5.2. Snapshots (a-d) correspond to the states at time moments $t_1 = 1500$, $t_2 = 3000$, $t_3 = 6000$ and $t_4 = 27900$ which are indicated by green dots on the trajectory. Both the front view (upper row) and the top view (bottom row) for all conformations are displayed. Absolute distances d_{12} in Å between the two motor domains at the respective time moments are indicated under the snapshots in the upper row. In the bottom row, the corresponding distances d_{13} are given. Coloring of the domains is the same as in Fig. 5.1. The backbone C_α-trace representation is employed.

deformed conformations of the network were obtained by integrating the equations of motion in the presence of random forces for a fixed time period.

5.1.1 Hef helicase

Figure 5.2 shows the relaxation pattern for the elastic network of Hef helicase. Grey trajectories start from different initial deformations obtained by applying global perturbations. Remarkably, no meta-stable states could be found. Even though many initial states corresponded to highly deformed conformations, the elastic network of the protein could always return to its equilibrium shape. Examining further the relaxation pattern, we notice that, while distances between the labels 1 and 2 or 1 and 3 could vary up to 40%, application of the same forces could change the distance between labels 2 and 3 by only a few percent. This suggests that the third domain in this protein is relatively rigidly attached to the second motor domain and they move essentially as a single object. Small initial changes u_{23} in the distance between motor domain 2 and domain 3 soon disappear and the relaxation trajectories become confined to a plane (see Fig. 5.2), on which subsequent slow relaxation takes place. We see that Hef helicase performs well-defined relaxation motions of the motor domain 1 with respect to domains 2 and 3, responding to random perturbation forces applied to all its residues.

Binding of an ATP ligand should produce local forces which are applied only to the residues in the binding pocket region. The exact details of such forces and of the induced responses in Hef helicase are not known. Nonetheless, we could generally probe the responses by applying various random static forces only to a subset of residues in the ATP binding pocket, similar to what we have done previously for the HCV helicase.

Relaxation processes starting from the states induced by the application of such local forces are shown as red trajectories in Fig. 5.2. One can see that the protein responds to the local perturbations in the ATP binding region essentially in the same way as to the generic globally distributed perturbations (grey trajectories). Well-defined relative motions of the motor domain 1 are again observed.

To further characterize domain motions we have produced a sequence of snapshots accompanying the conformational relaxation process that corresponded to one particular trajectory (highlighted black in Fig. 5.2). They are shown in the front and top views in Fig. 5.3. Large-amplitude hinge motions of the mobile motor domain 1 with respect to the other two domains, which are rigidly moving, are clearly seen. During conformational relaxation, the distance between the two motor domains increased from 37.9 Å to 44.5 Å so that a large-amplitude conformational change was observed. The distance between the first motor domain and the domain 3 was also increasing from 38.2 Å to 52.8 Å.

5.1.2   Hel308 helicase

By using the same methods, relaxation dynamics of the Hel308 elastic network has been investigated. Similar to Hef helicase, we found that the trajectories beginning from different initial deformations all returned to the equilibrium conformation with no meta-stable states present (Fig. 5.4). After short transients, trajectories converged to an attractive bundle along which the relaxation proceeded to the equilibrium confor-

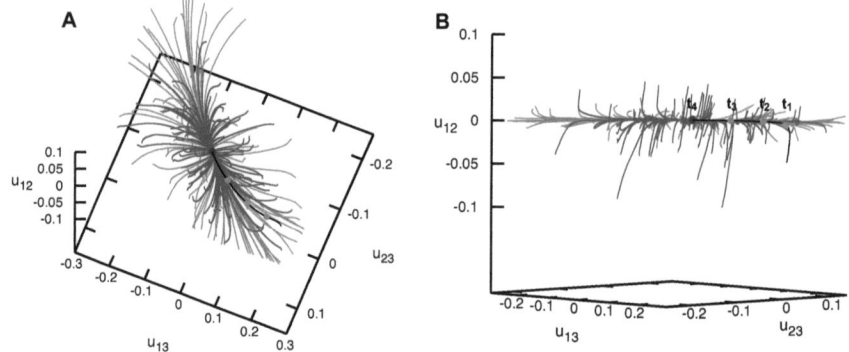

Figure 5.4: **Relaxation dynamics of Hel308 helicase.** Panels (A) and (B) show the relaxation pattern as a set of 100 trajectories in the space of normalized relative distance changes u_{12}, u_{23} and u_{13} between the three labels. Each trajectory starts from a different initial deformation that has been generated by applying random static forces globally distributed to all network beads (gray trajectories) or restricted to the beads of the ATP binding site (red trajectories). The viewpoint in (B) is chosen in such a way that the confinement of relaxation trajectories is best seen.

mation of the protein. This behavior was found for both sets of initial deformations, either obtained by applying random forces globally or with the random forces spatially confined to the ATP binding motifs.

In contrast to Hef helicase, the distance between the two motor domains, characterized by u_{12}, did not significantly change when perturbation forces were applied. On the other hand, large changes u_{13} and u_{23} in the distance between the third domain and the two motor domains could be induced by the same random forces, with the relative deformations reaching about 30%. Thus, the generic soft conformational dynamics in Hel308 helicase corresponds to large-amplitude ordered motions of the third domain with respect to the two motor domains. It is remarkable that large-amplitude motions of this domain can be generated by mechanical perturbations that are spatially confined to the remote ATP binding pocket.

To further illustrate the typical conformational relaxation process in Hel308 helicase, we show in Figure 5.5 a series of four snapshots at time moments $t_1 = 100$, $t_2 = 300$, $t_3 = 700$ and $t_4 = 5120$ taken along the trajectory, which is outlined in black in Fig. 5.4. One can clearly see that the top part of the third domain is mobile and can move substantially with respect to the two motor domains (with the distance changes

CHAPTER 5. CONFORMATIONAL MOTIONS IN SUPERFAMILY 2 HELICASES

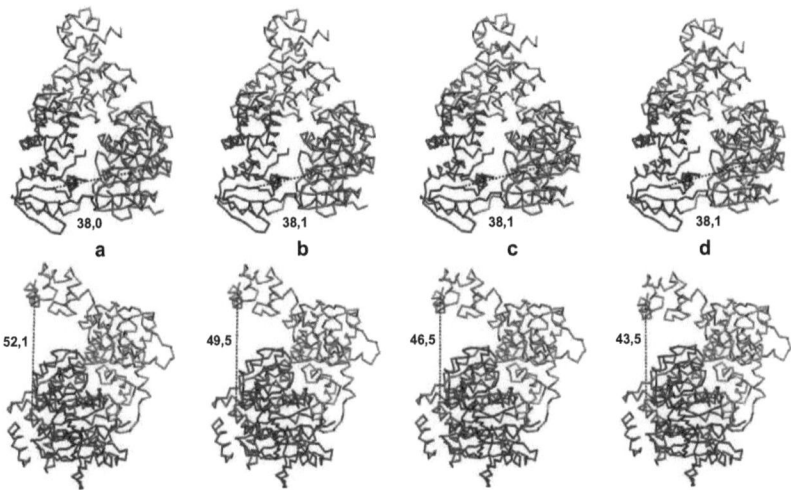

Figure 5.5: **Conformational motions in Hel308 helicase.** Consequent snapshots showing conformational changes in Hel308 helicase along the relaxation trajectory, highlighted in black in Fig. 5.4. Snapshots (a-d) correspond to the states at time moments $t_1 = 100$, $t_2 = 300$, $t_3 = 700$ and $t_4 = 5120$ which are indicated by green dots on the trajectory. Both the front view (upper row) and the side view (bottom row) for all conformations are displayed. Absolute distances d_{12} in Å between the two motor domains at the respective time moments are indicated under the snapshots in the upper row. In the bottom row, the corresponding distances d_{13} are given. Coloring of the domains is the same as in Fig. 5.1. The backbone C_α-trace representation for different conformations is used.

of about 9 Å).

5.1.3 XPD helicase

Finally, we present the results for XPD helicase, the last protein investigated by us. Relaxation processes starting from initial conformations generated by applying random forces to all network beads are shown as grey trajectories in Fig. 5.6. The initial induced deformations were strong, as evidenced by large relative distance changes between the labels (up to 40%). It can be moreover noted that, in this helicase, the perturbations induced pronounced rearrangements of all three domains. Nonetheless, the trajectories returned to the equilibrium conformation and meta-stable states were not found.

Relaxation trajectories starting from the second set of deformations, obtained by

5.1. *IN SILICO* INVESTIGATION OF SUPERFAMILY 2 HELICASES

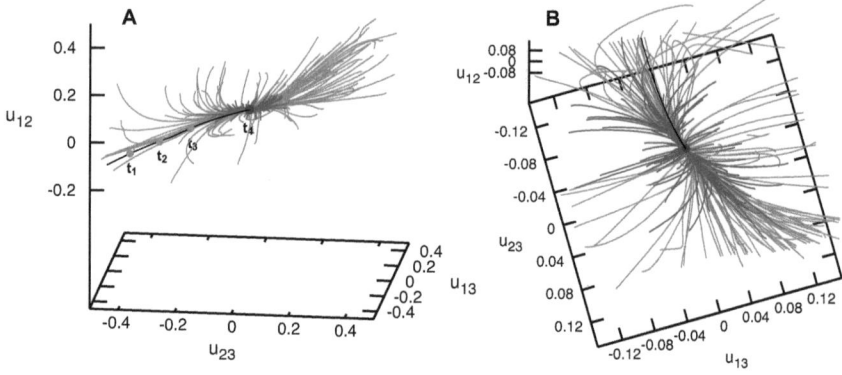

Figure 5.6: **Relaxation dynamics of XPD helicase.** Panels (A) and (B) show the relaxation pattern as a set of 100 trajectories in the space of normalized relative distance changes u_{12}, u_{23} and u_{13} between the three labels. Each trajectory starts from a different initial deformation that has been generated by applying random static forces globally distributed to all network beads (gray trajectories) or restricted to the beads of the ATP binding site (red trajectories). In panel (B), a different viewpoint is chosen and the scales are smaller.

applying random forces only to the residues in the ATP binding pocket, are shown as red curves in Fig. 5.6.

As we see, deformations induced by such spatially localized forces were significantly smaller, with relative distance changes between the labels not exceeding 10%. Even when stronger forces were applied, much larger deformations could not be found in our simulations (not shown in the figure). Thus, large conformational changes cannot be generated by mechanically perturbing residues of the ATP binding pocket only. This finding is in contrast to what we have observed for the two other helicase proteins and also for the previously studied HCV helicase.

Large-amplitude domain motions accompanying one particular relaxation process (corresponding to the black trajectory in Fig. 5.6) are shown as snapshots in the backbone-trace representation in Fig. 5.7. During conformational relaxation large changes in the relative positions of the motor domains can be seen (around 10 Å) and absolute changes in the distance between the third domain and the motor domain 2 are even larger.

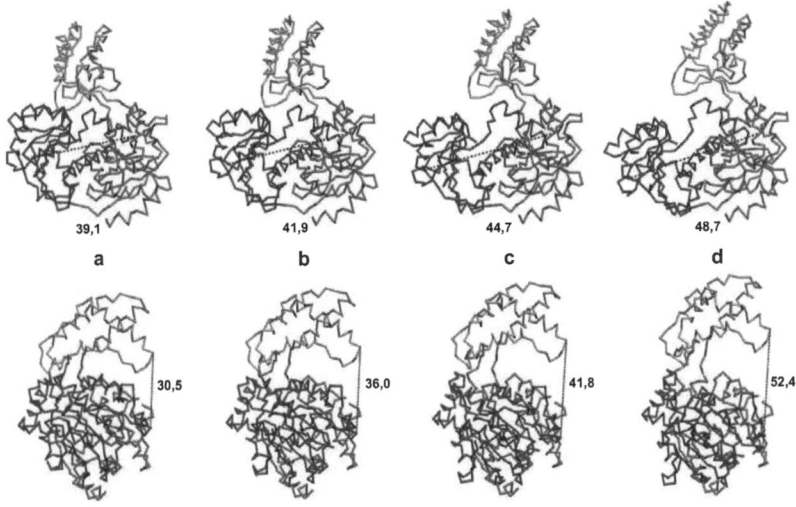

Figure 5.7: **Conformational motions in XPD helicase.** Consequent snapshots showing conformational changes in XPD helicase along the relaxation trajectory, highlighted in black in Fig. 5.6. Snapshots (a-d) correspond to the states at time moments $t_1 = 600$, $t_2 = 1200$, $t_3 = 2000$ and $t_4 = 9300$ which are indicated by green dots on the trajectory. Both the front view (upper row) and the side view (bottom row) for all conformations are displayed. Absolute distances d_{12} in Å between the two motor domains at the respective time moments are indicated under the snapshots in the upper row. In the bottom row, the corresponding distances d_{23} are given. Coloring of the domains is the same as in Fig. 5.1. The backbone C_α-trace representation for different conformations is used.

5.2 Functional aspects for helicase operation

The objective of our study presented in this chapter was to investigate conformational dynamics of proteins in the largest helicase superfamily within the framework of a coarse-grained mechanical description employing the relaxational elastic-network model. Three representative helicases of the superfamily 2 were chosen by us, i.e. the Hef, the Hel308 and the XPD helicase. To study conformational motions of the proteins, we have probed the dynamics of their corresponding elastic networks in response to various mechanical perturbations.

The common observed property of all studied helicases is that they respond in a well-defined way to mechanical perturbations and their responses consist in large-amplitude relative changes in the positions of the protein domains. Though the distances between the domains may change up to 40%, such conformational changes can still take place without protein unfolding, with the elongations of elastic links remaining relatively small. The relaxation starting from deformed initial conformations involves ordered motions of major protein domains. Metastable states could not be found and the molecules always returned to the original equilibrium conformation. Previously, a similar behavior has been observed for other motor proteins [118, 119] and it has also been found by us for the molecular motor HCV helicase.

The ability of motor proteins to perform ordered conformational motions triggered by ATP binding and hydrolysis is a fundamental property of molecular machines.

On the other hand, we have also seen important differences in the conformational dynamics of the three studied proteins. These differences can be important for the biological operation of the helicases, i.e. for the processing of the nucleic acid substrates. In the following the functional aspects are discussed separately for each of the studied molecule.

Hef helicase

The structure of this protein complexed with DNA is still not available. Nonetheless, well-known conserved motifs crucial for DNA binding have been identified in the two motor domains [148]. Analysis of the electrostatic surface potential and mutational studies of the protein have revealed that the third domain can recognize and bind specific (e.g. fork-structured) DNA. According to the model in Ref. [148], the two motor domains can bind double- or single-stranded DNA (see Fig. 5.8). Our study shows that in Hef helicase a mobile motor domain is able to perform large-amplitude hinge motions with respect to the other motor domain and the third domain, which themselves are rigidly connected. This observation is in accordance with what has been previously proposed based on crystallographic studies of this protein [148]. The hinge motion brings together or spatially separates, respectively, the residues of the conserved motifs which are located in the two motor domains. These motifs are involved in binding and hydrolysis of ATP and thus it is likely that such conformational motions are functional and essential for the motor operation. Our investigations suggest, that, in Hef helicase, DNA is actively transported by the two motor domains and becomes further processed

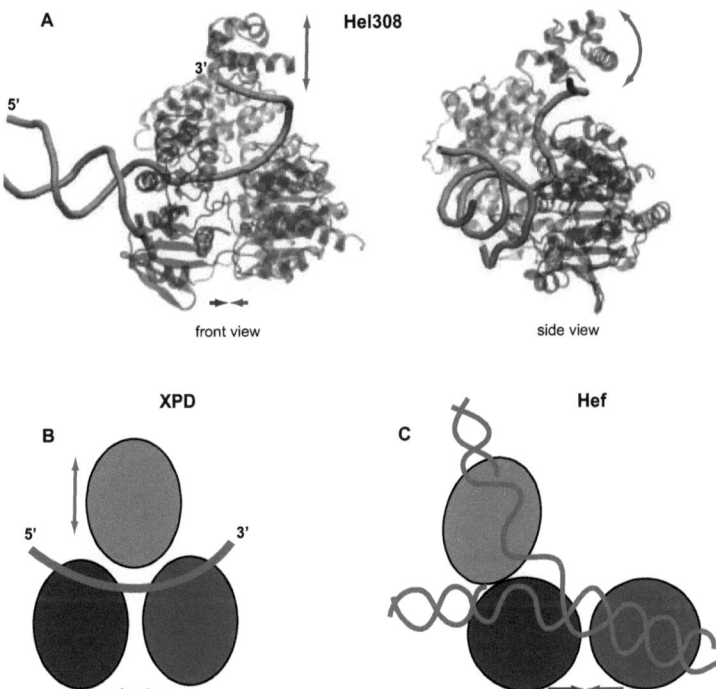

Figure 5.8: **Helicase function on DNA.** (A) Structure of Hel308 helicase with partially unwound DNA bound to it (2P6R) in the front and side views. The DNA strands are shown as green tubes; the direction of translocation along the unwound strand is 3' to 5'. The coloring of the domains is the same as in the other figures, but the top part of domain 3, referred to as the arch subdomain, is colored purple here. (B) Schematic drawing of XPD helicase with the single DNA translocation strand positioned according to Ref. [150]. In this protein, the direction of translocation is 5' to 3'. (C) Possible orientation of the branched duplex DNA in Hef helicase as proposed in Ref. [148] (schematic drawing). In all panels, arrows indicate possible domain motions.

by the third domain. This behavior closely resembles that of the HCV helicase, where large relative motions of the motor domains are enabling hydrolysis of ATP and drive progressive translocation along the nucleic acid. Remarkably, both in Hef and HCV helicase, the motions of the motor domains can be induced by forces applied only in the ATP binding region.

5.2. FUNCTIONAL ASPECTS FOR HELICASE OPERATION

Hel308 helicase

The structure of this protein with partially unwound DNA (PDB ID code 2P6R) reveals that the duplex DNA is bound to motor domain 2 and becomes separated by this domain [149]. The unwound 3' tail winds through the entire protein having contacts with all its domains. The backbone of the DNA strand goes across both motor domains, establishing contacts to the residues of the conserved nucleic acid binding motifs. Then it meanders towards the arch subdomain, binding also there as seen in Figure 5.8. The proposed operation mechanism of this helicase [149] consists of the processive helicase translocation in the 3'-to-5' direction accompanied by the ATP-dependent ratchet-like transport of the DNA strand, involving relative motions of the motor domains (inchworming). Mutational studies have revealed that the arch subdomain is of special importance for the helicase function. It has been found that the absence of this domain results in a significantly higher helicase processivity, as compared to the wild-type protein, and, therefore, an autoinhibitory or molecular brake role has been assigned to the arch subdomain [151]. Thus, it has been previously suggested that its function can be to limit and control the helicase activity. While our present investigations have shown that, in contrast to Hef helicase, the motor domains are already close one to another in the native conformation of this protein and their relative motions are less pronounced, the position of the DNA strand within the Hel308-DNA complex still suggests that such motions may be used for the ATP-dependent translocation. We have additionally found that the top part of the third domain, representing the arch subdomain [149], could perform large-amplitude hinge motions with respect to the two motor domains. Such motions could be induced both by globally distributed perturbations and by local forces applied only in the ATP binding pocket. As we have seen, conformational changes in the arch subdomain could be produced by perturbations acting in the ATP binding pocket, indicating correlations between the ATP binding or hydrolysis and the regulation of helicase activity and thus the presence of long-range internal communication in this protein. Such possible correlations have previously been also discussed based on biochemical analysis of the Hel308 helicase [152]. Our simulation results suggest that the arch subdomain does not merely function as a passive brake, but has an active functional role. It can mechanically regulate the grip on the 3' DNA tail in an

ATP-dependent fashion. Thereby, the arch subdomain may actually operate as a clamp to control the activity on the DNA substrate.

XPD helicase

Structural data for the XPD helicase-nucleic acid complex is not yet available. Nonetheless, the conserved residue motifs, known to interact with the nucleic acid substrate in other helicases, have been identified in this protein [150]. They are located on the surface of the two motor domains. Using this data, the single-strand of DNA can be positioned like in the well-resolved Hel308-DNA complex [150] (see Fig. 5.8). We have found that in XPD helicase significant changes in relative positions of all three domains accompany conformational dynamics. However, in contrast to the two other helicases and HCV helicase, they could not be generated by mechanically perturbing residues only in the ATP binding site. Thus, additional mechanical stimuli, provided e.g. by interactions with the nucleic acid substrate, may be needed in this protein to yield significant motions of the motor domains.

For a different helicase in the same superfamily 2, it could indeed have been shown that binding of the nucleic acid to the apo structure may induce large repositioning of the motor domains, thus affecting the binding affinity of ATP molecules to the protein and controlling the ATP-dependent helicase activity [153]. We cannot however also exclude a possibility that the conserved residue motifs, which have so far been identified for XPD helicase, do not actually account for all residues constituting the ATP binding pocket. It may therefore turn out that large-scale conformational changes may still be induced by local perturbations in this pocket, but the forces should then be applied additionally to some other residues. Indeed, in the crystal structure used by us, residues of the conserved ATP binding motif V could not be determined [150] and were therefore missing in our simulations.

Generally, one can expect that the dynamical behavior of helicases, regarded as a result of biological evolution and selection, should be adapted to the particular molecular processes in which they are involved and thus to the specific forms of the nucleic acid substrate. In viruses, where helicase are functioning in the molecular replication machinery together with polymerase proteins, they need to translocate over large distances along the nucleic acid strands and separate their duplex structure, in order to ensure efficient multiplication of the viral genome.

The situation is different when branched forms of nucleic acids should be separated during transcription processes or intermediate bubbles in the regular duplex structures should be generated in the processes of DNA repair. In this case, helicases need to catalyze only local unwinding of short duplexes and aspects of efficient translocation might be less important. Instead, the execution of specific functions, such as nucleic acid recognition and fine-regulated processing of nucleic acids, can then play a dominant role in conformational dynamics.

Two possible operation mechanisms have previously been discussed for superfamily 2 helicases [60, 154]. According to the active mechanism, a helicase can progressively move itself along the nucleic acid substrate, thereby destabilizing its duplex structure and eventually separating the strands. The passive mechanism assumes instead that, once a base pair becomes broken due to thermal melting, the helicase passively moves along the strand and locally locks the destabilized region. In our study of HCV helicase, we could investigate entire operation cycles of active nucleic acid unwinding, with the motor domains acting as a translocation machine. In that study, interactions with the DNA strands have been explicitly included into the considered dynamical model. For the three presently chosen helicases, we could not yet consider entire operation cycles, including interactions with nucleic acids and unwinding processes. We can, however, note that when large large relative motions of the two motor domains are observed, this may indicate the active mode of operation including stepwise translocation along the nucleic acid strand, though further investigation is needed to clarify the situation.

5.3 Simulation details

To study conformational dynamics of superfamily 2 helicases we have used methods that were similar to those applied to the HCV helicase. The dynamical responses of the chosen helicases to deformations of their elastic networks were analyzed in order to study conformational motions and clarify their functional aspects. In the same way as for the previously considered HCV helicase, we needed to generate various deformed states by applying random static forces to particles of the network. Again we have distinguished between deformations that have been obtained by applying the forces globally distributed over all network particles, or only to a subset of particles that corresponded to residues of the known ATP-binding pocket in the protein. In each realization of random forces the normalization condition $(\sum_i |f_i|^2)^{1/2} = C$ was satisfied. To obtain the coordinates of particles in the deformed state, the set of dynamical equations (2.9) was numerically integrated in the presence of the forces for a fixed time t_f. This procedure was repeated to prepare a set of 100 initial deformations, each arising from a different random configuration of forces. For each prepared initial deformation, we have checked that the springs were not overstretched, i.e. that plastic deformations were excluded. To imply this, we have required that elongations of the springs do not exceed $1.5 \cdot d_{int}$. When global deformations were considered, numerical values used in the simulations were $C = 1.0$ Å and $t_f = 10000$ for Hef helicase, $C = 5.0$ Å and $t_f = 20000$ for Hel308 helicase and $C = \sqrt{3.0}$ Å and $t_f = 10000$ for XPD helicase.

Since none of the chosen helicases have been co-crystallized with an ATP-analog yet, details of their interactions are not known. Nonetheless, the conserved residue motifs which have been identified in the helicases to be involved in ATP binding and its hydrolysis [148, 149, 150] have been used in our simulations. Assuming that interactions between ATP and the helicase produce local forces in the ATP binding region, we have probed helicase dynamics in response to mechanical perturbations confined to the ATP binding pocket and using the aforementioned binding motifs in each of the helicases.

The following residue motifs were used in our simulations. For the Hef helicase: Thr^{32}-Gly-Leu-Gly-Lys-Thr (motif I), Val^{130}-Phe-Asp-Glu-Ala-His (motif II) and Glu^{455}-Arg-Arg-Gly-Arg-Thr-Gly-Arg (motif VI). For the Hel308 helicase: Gly^{52}-Lys-Thr-Leu-Leu (motif I), Asp^{145}-Glu-Ile-His-Leu (motif II) and Gln^{365}-Met-Ala-Gly-Arg-Ala-Gly-Arg-Pro (motif VI). For the XPD helicase: Gly^{36}-Leu-Gly-Lys-Thr (motif I), Asp^{182}-Glu-Ala-His (motif II), Ser^{326}-Gly-Thr (motif III) and Gln^{500}-Thr-Ile-Gly-Arg-Ala -Phe-Arg (motif VI). In these simulations, the values $C = \sqrt{3.0}$ Å for Hef helicase, $C = 10.0$ Å for Hel308 helicase and $C = \sqrt{5.0}$ Å for XPD helicase have been used.

The values for t_f were the same as before.

The visualization of conformational motions has been carried out in the same way as done previously for HCV helicase. Labels 1 and 2 were chosen to lie in the motor domains 1 and 2, whereas label 3 belonged to the third domain. For Hef helicase the labels 1, 2 and 3 were Ser^{91}, Ala^{374} and Lys^{264}. For Hel308 helicase we have taken Val^{62}, Ile^{382} and Val^{679}. For XPD helicase we have chosen Ala^{140}, Thr^{427} and Glu^{273}.

Chapter 6

Models of synthetic protein motors

So far we have considered the dynamics of real protein motors. The dynamical systems investigated in this chapter are of different nature. Here, we are considering dynamical properties of artificially constructed analogs of real biological objects. We explain the construction of a model artificial motor and investigate its behavior in the presence of thermal fluctuations.

6.1 Introduction

Our investigations of the HCV helicase motor protein and other representative motors from the largest helicase family have shown that such proteins share one common property. They can perform highly coordinated ordered conformational motions that are robust against perturbations and exhibit only a weak dependence on the initial conditions. This property is fundamental as it underlies the ability of molecular machines to reproduce the same coordinated internal movements in each next operation cycle.

To understand the fundamental operation principles of molecular machines it is very important to study the proteins in experiments under real biological conditions. Since this turns out to be complicated and often is not possible at all, it can be on the other side also advantageous to investigate simplified model systems under controllable conditions in theoretical studies. Well-known theoretical descriptions for motor proteins are provided by the ratchet models in which their dynamics is drastically reduced (see section 1.4). Since they are lacking any structural aspects of the proteins, it would be desirable to construct models systems that go beyond this over-simplified picture and

6.1. INTRODUCTION

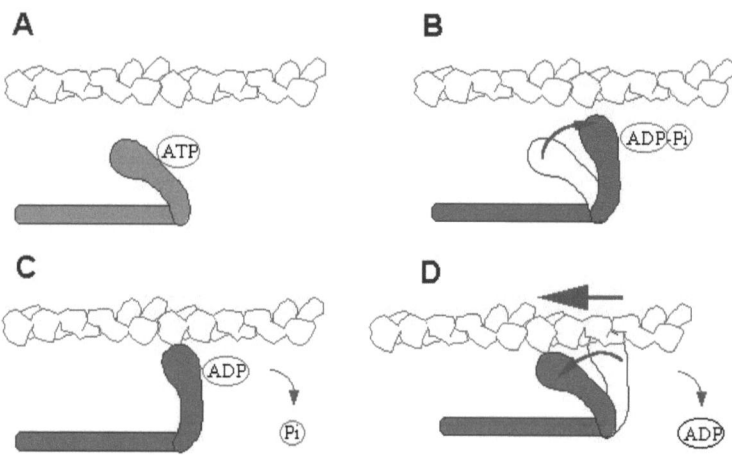

Figure 6.1: **Muscle-myosin motor.** A single-head myosin motor (blue) can interact with filamentous actin (red shape) and generate a force on it under its ATP-dependent activity [1].

allow to resolve also structural changes accompanying protein operation.

It is also clear that due to the enormous complexity it is not yet possible to create models that take into account complicated details of real motor proteins and mimic their operation with all complex aspects.

Therefore, in this chapter we want to engineer model systems of intermediate complexity based on the elastic network description. They were intended to incorporate important dynamical aspects of real biological motor proteins in a simplified fashion.

In a previous work it has been shown already that the elastic network description can be used to construct simple model systems of molecular machines [118]. There, elastic networks with special dynamical properties have been designed and a prototype machine was constructed. Powered by energy supplied with a ligand, they were able to undergo ordered cyclic structural changes.

The process which will be used by us to design special networks relies on the methods similar to those of the previous work [118]. However, in contrast to [118], we shall go further and construct a model of a synthetic protein motor.

While the term 'protein machines' generally refers to any nanoscale device that

---

[1] Taken from the Maciver Lab webpage (http://www.bms.ed.ac.uk/research/others/smaciver/).

cyclically changes its structure upon binding of a substrate ligand, the operation of molecular motors is more specific. In the motors, cyclic internal changes are used to produce a force on other molecules. In the kinesin and dynein molecules, which represent well-known motor proteins, internal changes of their conformation induced by binding and hydrolysis of ATP molecules are translated into mechanical work applied to microtubules. In this way these motors can transport molecular cargo through the cell by walking along the microtubules [17].

Our aim was to consider a model synthetic motor that implements features of real motor proteins. We have therefore constructed a setup consisting of an artificial machine and an artificial filament. A ratchet mechanism was implemented, allowing for the transport of the filament by the machine. This model system is used for further investigations of the principles of protein motor operation.

The artificial motor constructed by us is roughly analogous to the muscle myosin, a protein motor that interacts with actin filaments to control functions in muscle cells (see Fig. 6.1).

Additionally, we have included into the model the noise accounting for thermal fluctuations into the dynamical description.

6.2 Design of elastic networks

First, we describe how elastic networks with special dynamical properties have been designed.

The design principles for the elastic networks basically rely on the application of optimization methods intended to improve and eventually perfect the performance of a network, which is subject to continuous evolution, according to the prescribed properties. There are at least two key questions here. The first is how one should organize the evolution process of a network and the second is what kind of criteria one should employ for the optimization.

In the real biological situation, evolution of proteins and their functions proceeds through point mutations in the amino acid sequence, i.e. the local replacement of an amino acid by another one, which can result in different protein folds and thus different

6.2. DESIGN OF ELASTIC NETWORKS

functional properties. The decision which fold is better than another one is then, simply speaking, determined by the evolutionary pressure a protein is subject to.

It is clear that this scenario is too complex to be implemented in computer simulations. In order to keep the computational effort bounded, Togashi and Mikhailov have considered the following situation [118]. Starting from a random elastic network, subsequent single structural mutations that can locally change the interaction pattern are performed. To evaluate the worth of evolving structures, the spectral gap of a network before a structural mutation has been derived and compared with that after the mutation.

Although the dynamics of an elastic network is generally nonlinear (see section 2.3), properties of the linearized version of the respective network, namely its eigenvalue spectrum, have been employed for the selection method. This particular choice turned out to be successful in order to design networks with coordinated well-defined motions.

According to the linearized model, the over-damped relaxation dynamics of the elastic network is governed by exponentially decaying normal modes with the lifetimes given by the inverse of the respective eigenvalues. Let us now imagine the following situation: We consider two networks with two different spectral properties. Suppose that in one network we find a gap in the spectrum that separates a few lowest eigenvalues from the rest of the spectrum, while in the other the eigenvalues are very densely distributed and no significant gap can be identified. Now, when initial structural perturbations from the equilibrium conformation of the network are considered, the latter case with absent gap implies that even after long times a large number of relaxation modes will contribute to the dynamics. In the other case instead, where a significant spectral gap is present, all modes with large eigenvalues beyond the gap will quickly die out and the long-term relaxation will be determined by a few (or even a single) relaxation modes. When relaxation dynamics of the elastic network is dominated by a few slow modes, the corresponding motions will be accordingly simple (i.e. low-dimensional) and can be described in terms of a few mechanical coordinates only.

It turns out, that when spectral properties of elastic networks corresponding to real biological motors are considered, a significant spectral gap can often be found [118, 119]. The evolutionary optimization method applied by Togashi was therefore aimed

CHAPTER 6. MODELS OF SYNTHETIC PROTEIN MOTORS

to design networks with a substantial spectral gap.

Since similar design methods for the networks are also employed by us, we explain the single steps underlying the design process in greater detail. The first step consists in the construction of a random elastic network with which the later evolution is performed. To do this, a chain of N identical particles is randomly folded in the three-dimensional space. After the first particle has been placed, the second and each next particle are randomly positioned under the condition that their distance l from the previous particle comes from an interval $[l_{min}, l_{max}]$. Here, to avoid a too compact chain fold, each placed particle is forced to have at least the distance l_{min} from all previous particles. The randomly folded chain of identical particles represents the backbone of the network. Under the folding process different particles have presumably come close one to another and the construction of the random network is now completed by additionally connecting those pairs of particles with a spring that have a distance shorter than a prescribed interaction radius l_{int}. We have assumed that $l_{int} > l_{max}$.

Now the evolutionary optimization can be undertaken. It consists of subsequent events of structural mutations followed by selection. A mutation is carried out by changing the network structure locally. To do this, a particle is chosen at random and its equilibrium position is replaced by a new equilibrium position which is randomly chosen within a sphere of radius $d = l_{min}$ around the old one. A side condition aimed to preserve the backbone geometry is implied in such a way that we require both distances to the neighbor particles in the chain to be still within the interval $[l_{min}, l_{max}]$. Furthermore we have excluded the possibility that other pair distances become shorter than l_{min}. The new pattern of connections for the mutated particle is determined by examining distances to all other nodes and creating spring connections if they are below the interaction radius l_{int}.

For the selection process, that is a decision whether a mutation was good or bad, the spectral gap before the mutation and after it was examined. The gap was defined as the logarithm of the ratio between the two lowest spectral eigenvalues λ_1 and λ_2, i.e. $g = \log_{10}(\lambda_2/\lambda_1)$. If the gap g' after a mutation was larger than that before the mutation g, i.e. $\Delta g = g' - g > 0$, the mutation was always accepted. If instead the gap had decreased under the mutation, i.e. $\Delta g < 0$, we have still accepted the mutation albeit with a small probability $P = \exp(\Delta g/\theta)$ (θ being the optimization temperature) and otherwise (i.e. in most cases) rejected it.

The latter condition prevents that the system gets stuck in an intermediate optimal configuration which may not coincide with the global optimum of the network.

Evolution of the network is obtained by iteratively applying the sequence of mutations followed by selection. The method of evolutionary optimization employed here is reminiscent of the Metropolis algorithm which is widely used in optimization problems.

After a fixed number of evolutionary steps, networks with a large spectral gap have been obtained. Then, as in [118], a few steps of neutral evolution have been performed. This means that several subsequent mutations have been performed without selection. Under this process the spectral gap was typically decreasing.

It turns out that many of the designed elastic network have a modular structure and indeed obey the desired dynamical properties. Their behavior is fundamentally different from the random networks. When exposed to initial deformations, the random networks behave similar to amorphous glassy systems and possess many metastable conformations different from the true equilibrium state. The designed networks instead respond by performing well-defined organized relaxation motions that lead back to the original equilibrium conformation of the network (see [118]). It is highly remarkable that, although properties of the linearized elastic network have been used for the optimization routine, for strong deviations from the equilibrium conformation, where the linear approximation should definitely fail, the network dynamics is still low-dimensional and can be described in terms of simple domain motions.

The evolutionary optimization of elastic networks was implemented by the following procedure. We have prepared a set of 200 initial random networks and have ensured that they do not possess degrees of freedom additional to translational and rotational invariance, i.e. their spectrum had to have $3N - 6$ non-zero eigenvalues. In each evolution, a fixed number of $3,500$ steps was performed. The numerical values $l_{min} = 3.0$, $l_{max} = 4.0$, $l_{int} = 7.5$ and $\theta = 0.1$ were used in the simulation.

After that, networks with a gap larger than 9.0 have been taken and further inspected. Out attention was focused on the structure and flexibility properties. In each network, the links can be colored according to the deformation they would undergo when network motion in the lowest eigenmode is considered (see [118]). This provides information about the network flexibility. As we have observed, many of the designed networks consisted of rigid parts connected by a flexible region, indicating that the rigid parts may undergo large relative motions with respect to each other. We have therefore selected some of the designed networks in which the structure was separated into two rigid bodies connected by a hinge region. For each of the networks, we have then performed a few steps of neutral evolution to decrease the spectral gap. Here, we have imposed the condition that this process should not generate additional zeros in the spectrum.

Finally, after performing the described selection, we have probed the mechanical properties of the few chosen designed networks. This was done in the same way as for the real proteins, i.e. by monitoring the relaxation trajectories of an elastic network in response to different initial deformations.

Our primary aim at this point of the study was to cast a single candidate network that can be later used as a template for the synthetic motor setup. We were not therefore collecting any statistics that can provide information about the efficiency of the design process. In the previous work by Togashi, more attention has been put on the statistical aspects and such data has been then indeed collected [118].

Several designed networks have shown well-defined robust relaxation properties similar to those found for elastic networks of real proteins. We have chosen one specific network around which further investigations will be centered.

6.3 Designed network template

The designed network chosen by us for further investigations contains $N = 50$ particles and with the prescribed interactions radius of $l_{int} = 7.5$ gives rise to 240 spring connections. After the entire optimization process including the neutral evolution its spectral gap is $g = 1.1$. The architecture of this network is composed of two rigid domains that are connected flexibly by the middle part (see Fig. 6.2). The dynamical properties of this network have been probed by testing its relaxation response to various initial de-

6.3. DESIGNED NETWORK TEMPLATE

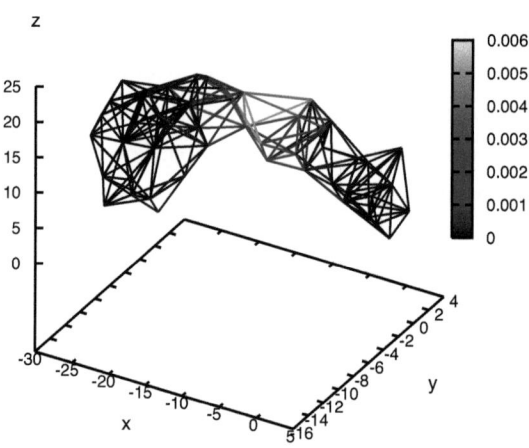

Figure 6.2: **Designed network template.** The designed network with the links colored according to their relative deformations in the motion corresponding to the slowest normal relaxation mode (for the method see [118]).

formations which have been generated by applying random static forces to all network particles. The procedure follows exactly that used for the investigations of the HCV helicase and the other helicase motor proteins and, to avoid repetitions, we refer to earlier parts of this thesis for detailed explanations (see section 3.2 [2]). The difference was that only initial deformations obtained by perturbing the network globally have been considered.

Conformational relaxation motions of this network have been visualized by selecting three particles as labels and recording distance changes between them. We have placed two labels in one part of the network and the third label inside the distant other domain (see also [118] for a routine to chose the labels). The chosen labels 1, 2 and 3 correspond to the network particles with indices 47, 10 and 0.

The network and its relaxation pattern are shown in Fig. 6.3. We see that by the

---

[2]In the present simulation the numerical values $C = 1.0$ and $t_f = 10,000$ were used.

CHAPTER 6. MODELS OF SYNTHETIC PROTEIN MOTORS

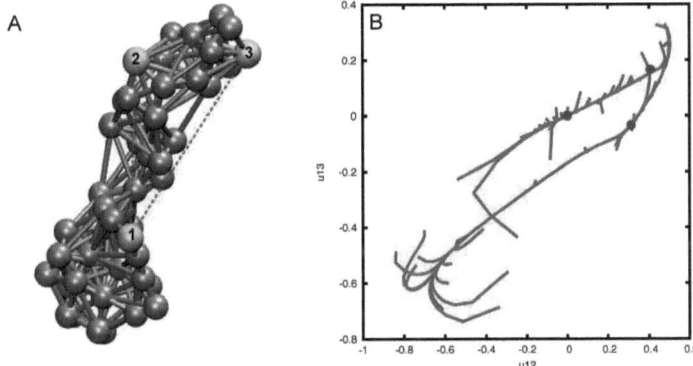

Figure 6.3: **Designed network with robust relaxation dynamics.** (A) The designed network consisting of 50 particles and 240 links is shown. Labels 1, 2 and 3 used for monitoring conformational changes are highlighted as yellow balls. (B) Red lines show relaxation trajectories in the plane of normalized distance changes u_{12} and u_{13} between the chosen labels, each starting from a different initial conformation obtained by applying random static forces to all particles.

application of random forces this network can undergo large deformations with relative changes of up to 60% between the domains. The relaxation trajectories starting from those highly deformed states proceed along an attractive path and the dynamics corresponds to organized relative motions between the two domains.

In most cases, the network is able to return to its original equilibrium conformation. There is, however, also one meta-stable present, indicating that the elastic energy landscape possesses (at least) a second deep energy minima. This state is located far enough from the true equilibrium so that no impact needs to be expected when further cyclic ligand-induced motions of this network will be considered. As evidenced by the relaxation pattern, conformational motions of this network consist in large-amplitude hinge motions of the two domains with respect to each other. The flexible middle connector part serves as a mechanical hinge.

6.4 Ligand-induced cyclic operation

In the following, we will explain how the designed network is used to construct a cyclic operating synthetic machine powered by binding of a fictitious substrate ligand and its

6.4. LIGAND-INDUCED CYCLIC OPERATION

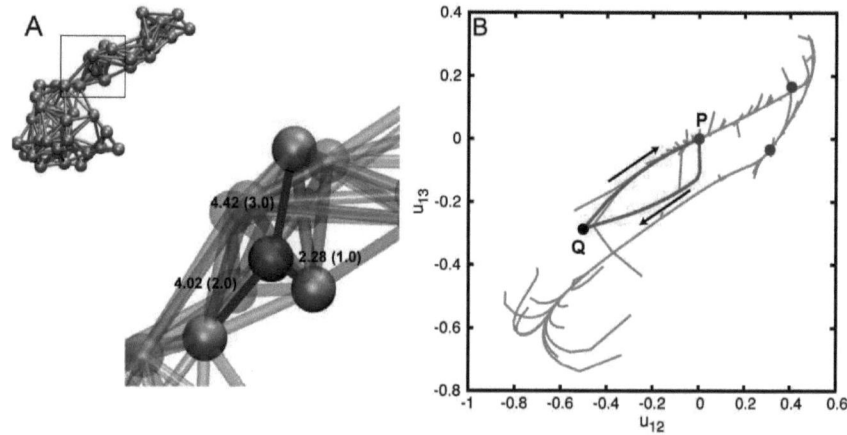

Figure 6.4: **Substrate binding and cyclic relaxation.** (A) The substrate particle (red) sphere can establish elastic links (shown in blue) to three binding site particles (gold spheres) in the flexible hinge region. Initial lengths of the additional links and their respective prescribed natural lengths (in parentheses) are given. (B) The conformational change P → Q is induced upon binding of the substrate ligand particle; the reverse motion Q → P is observed when the ligand, after its conversion from substrate into product, is released. For comparison, gray lines show the relaxation trajectories (the same as in Fig. 6.3).

conversion. The technique applied here is similar to that used for modeling of ATP-induced cyclic changes in HCV helicase.

There, a substrate ligand particle has been placed into the ATP binding region bringing energy through newly formed deformed elastic links. This induced conformational motions of the network-substrate complex towards its equilibrium conformation. At this stage, a fictitious conversion of the substrate particle into the product particle was assumed. The product had no connections to the network and was therefore released. Subsequent back relaxation to the equilibrium of the free network then proceeded.

Since we are dealing with an artificial network, no natural binding pocket for the substrate is present. Instead, we define three interaction sites for the substrate located in the flexible hinge region of our designed network. The substrate particle is placed in the geometric center of these binding site particles and interactions are mediated by newly formed links to them. Those additional links are initially deformed so that the creation of the network-substrate complex upon substrate binding is accompanied by

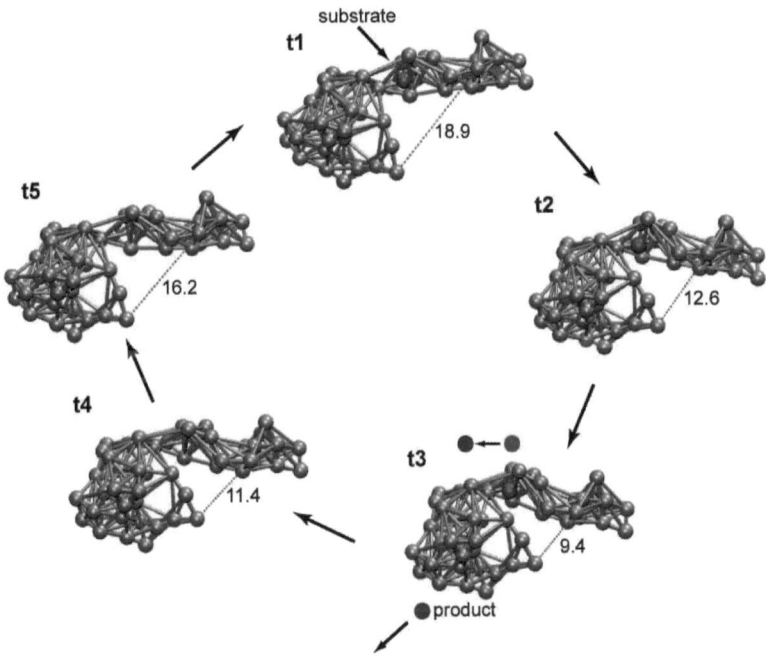

Figure 6.5: **Cyclic operation of a synthetic machine.** Binding of the substrate particle (red sphere) to the equilibrium state of the network at time moment $t_1 = 0$ induces large-amplitude domain motions inside the network-substrate complex until its equilibrium conformation is reached at $t_3 = 30,000$. The product particle (blue sphere) is ejected and the free network performs relaxation motions towards its original equilibrium conformation. Conformational snapshots at time moments $t_4 = 30,100$ and $t_5 = 30,400$ are shown. In all snapshots, absolute distances between the labels 1 and 2 are indicated.

a supply of energy. When the links are established, we assign to each of them some natural length that is smaller than the respective initial one, meaning that they are initial stretched and interactions with the substrate ligand are of attractive kind. The same method has been applied previously when a ligand-powered machine was constructed [118].

The choice of the substrate interaction sites within the flexible network region is to some extend arbitrary. Our aim was to construct a machine that is able to perform substantial structural rearrangements upon binding of the fictitious substrate particle. It is obvious that many different configurations are possible, and, to keep it simple we have just tried various of them including different binding sites as well as different natural lengths of the three additional links in order to find the largest response in terms of large-scale domain motions. The finally chosen binding configuration is shown in Fig. 6.4. They correspond to network particles with indices 17, 19 and 21.

The substrate particle that was bound in the geometric center of this pocket has created additional links of lengths $(2.28, 4.02, 4.42)$. The assigned natural lengths of the links were $(1.0, 2.0, 3.0)$.

With this choice of parameters, the machine could undergo substantial motions with relative distance changes between the domains of up to 50% under binding of the substrate ligand (see Fig. 6.4). When the ligand, after its conversion from substrate into product, is released, the network can return to its original equilibrium conformation along a relaxation trajectory that lies within the attractive bundle (see Fig. 6.4). From Fig. 6.4, we also see that the meta-stable state will not be involved in the cyclic operation of the artificial machine.

Snapshots from a video displaying conformational changes within one operation cycle of the machine are shown in Fig. 6.5. Upon binding of the substrate, a large-amplitude hinge motion brings together the two domain parts of the network and after ligand removal at time moment $t = 30,000$, they become separated again. One cycle is finished after the time $t = 60,000$.

6.5 A prototype synthetic protein motor

What we have in hand now is an artificial counterpart of real existing molecular machines. It is able to perform cyclic well-organized conformational motions powered by discrete supply of energy brought by a fictitious ligand particle. We have further considered a setup of the machine and an artificial filament. By applying further extensions to the model step-by-step, we were eventually aiming to establish a model system that, although within the limitations of the approximate description, was able to mimic the activity of molecular motors and thus made the investigation of mechanical aspects of their operation principles possible.

CHAPTER 6. MODELS OF SYNTHETIC PROTEIN MOTORS

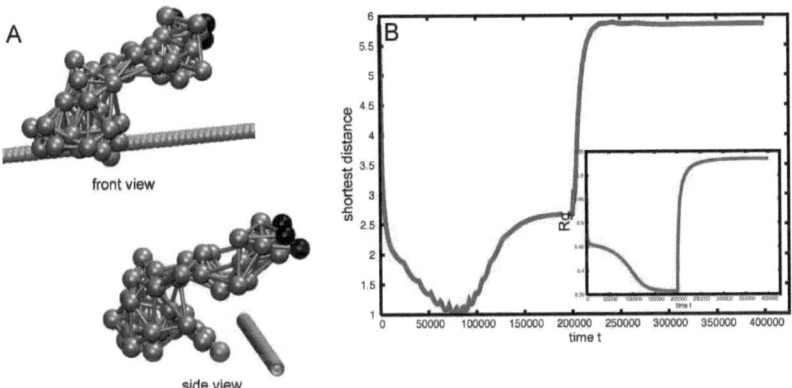

Figure 6.6: **Prototype machine with filament.** (A) The machine-filament setup shown in the front and side view perspective. The particle that can make contact to the filament is highlighted in orange color and three immobilized particles from the tail domain are shown in black. (B) The shortest distance of the machine's orange particle to the filament is plotted as a function of time during one ligand-induced cycle. The inlet shows the change in the radius of gyration of the substrate binding pocket.

6.5.1 Interactions with an artificial filament

Our aim was to demonstrate how ligand-induced internal motions of the synthetic machine can be translated into the directional movement of a filament. When the machine-filament complex is considered below, one machine domain will be allowed to connect to the filament. It will be referred to as the head domain. The other domain will be referred to as the tail domain.

The artificial filament is modeled as a chain of identical beads that is placed below the head domain of the machine. The entire setup can be seen in Fig. 6.6 in two different perspectives. Having in mind applications that include the transport of this filament by the machine, we needed to clarify two important aspects. First, we had to specify interaction sites between the machine and the filament. To keep it simple, we have chosen only one particle of the machine's head domain for a potential interaction spot, the particle with index 47. Second, we needed to immobilized the machine at the top part of the tail domain. The immobilized particles were those with indices $0, 5$ and

6.

With an appropriate placement of the filament, we have now modeled one ligand-induced operation cycle of the artificial machine and measured the shortest distance of the selected interaction particle of the head domain to the beads of the filament. During this cycle, no interactions between the machine and the filament have been included yet. The filament was also static. We just needed to check whether the relative arrangement of the machine and the filament was properly implemented and to specify the parameters needed for further simulations that actually imply interactions with the filament. The results are seen in Fig. 6.6. After binding of the substrate ligand, the head domain moves towards the filament reaching a minimal distance around 1.0. Upon removal of the ligand the head domain rapidly rotates away from the filament and under subsequent conformational relaxation back to the equilibrium state of the machine, it moves parallel to the filament maintaining a constant distance around 6.0.

Thus, the observation shows that the machine can come close to the filament when the substrate ligand is present and remain at a distance from the filament in the absence of the ligand. A ratchet mechanism for the filament transport by the machine can be thus implemented when distance-dependent physical interactions between the machine and the filament are introduced.

To do this, we have prescribed a distance threshold that controls the binding affinity of the head domain to the filament. Once this threshold is crossed, we assume the affinity to be suddenly changed. If the shortest distance between the head domain and the filament falls below the threshold, its affinity becomes very large and an additional link between the binding particle of the head domain of the machine and the filament becomes established. When the threshold is exceeded, it is assumed that the binding affinity vanishes and therefore the additional link becomes removed. When the link is present, the machine will exert a force on the filament and it will follow the motion of the head domain. In our simulations, we have assumed for simplicity that the filament is absolutely rigid (no internal deformations are allowed) and its motion is confined to the direction that coincides with its initial orientation.

In the simulation, we have set the threshold for binding/detachment of the machine's head domain to the filament at 1.5. The additional link mediating the interaction be-

CHAPTER 6. MODELS OF SYNTHETIC PROTEIN MOTORS

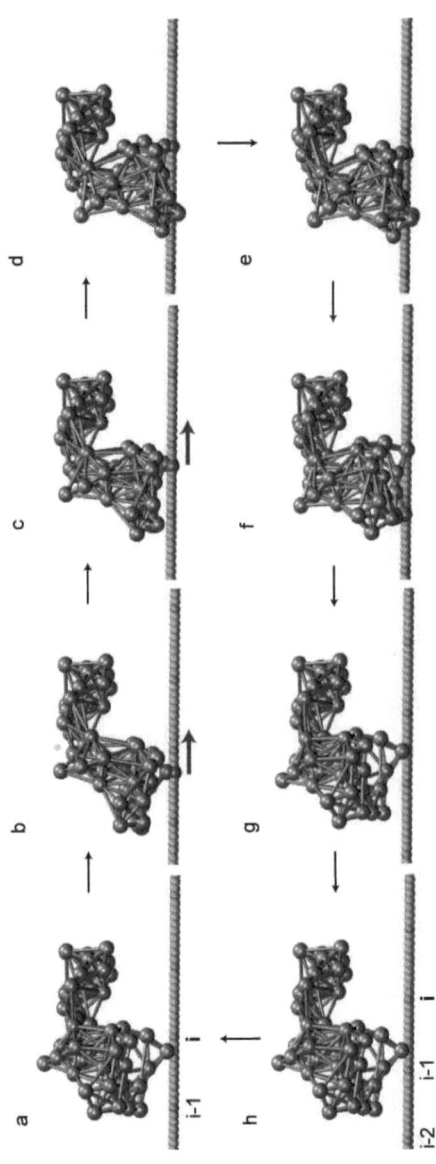

Figure 6.7: **Motor operation on a filament.** (a-h) Snapshots from a simulation of one machine operation cycle including interactions with the filament at different subsequent time moments. After binding of the substrate ligand the machine can bind to the filament and produce a force on it (b-c). When the equilibrium state of the network-substrate-filament state is reached (d) and the ligand is removed, the head domain will disconnect from the filament (e-f) and the machine will move back to its initial conformation (h). The effective transportation of the filament under the operation cycle can be best seen by comparing snapshots (a) and (h).

93

tween the machine and the filament is described by us in terms of the Morse potential $D[\exp\{-\alpha(d_{mf} - d_{mf}^{(equ)})\}]$, with the parameters $D/\kappa = 0.01$ and $\alpha = 3.0$. Here, d_{mf} is the actual length of the link to the filament and $d_{mf}^{(eq)} = 1.0$ is its prescribed equilibrium length.

We have run a simulation of a single operation cycle including interactions of the machine with the filament. Snapshots of this simulations are shown in Fig. 6.7. After binding of the substrate ligand, the machine is able to bind to the filament and thus switches into its force-producing state, in which the head domain can drag the filament into one direction.

Through this coupling, the internal motions of the machine are translated into directional motion of the filament. By the time moment when the equilibrium state of the machine-filament complex is reached (part (d) in Fig. 6.7 at time moment t=200,000), the immediate conversion of the substrate into the product particle and its release leads to the decoupling of the machine from the filament. The machine alone can then return to its initial conformation without carrying with it the filament. During one machine operation cycle, the filament therefore becomes effectively transported into one direction (compare (a) and (h) in Fig. 6.7).

6.5.2 Motor operation under thermal fluctuations

In the previous investigations of the HCV helicase and other superfamily 2 helicases we have neglected the effect of noise on the dynamics in the simulations. In their natural environment, however, the proteins are always subject to thermal fluctuations which can be also quite strong.

For the discussed model system of a protein motor, we therefore wanted to include effects of noise as well.

Noise accounting for thermal fluctuations can be imbedded into the present dynamical elastic network model by assuming that network particles are subject to additional time-dependent random forces with certain intensities. These external forces can be

CHAPTER 6. MODELS OF SYNTHETIC PROTEIN MOTORS

added to the equations of motions (2.9). Explicitly they read

$$\dot{\vec{R}}_i = -\sum_{j=1}^{N} A_{ij}(|\vec{R}_i - \vec{R}_j| - |\vec{R}_i^{(0)} - \vec{R}_j^{(0)}|)\frac{\vec{R}_i - \vec{R}_j}{|\vec{R}_i - \vec{R}_j|} + \vec{f}_i \qquad (6.1)$$

The time dependent random forces $\vec{f}_i$ acting on particle i had the intensity σ and their components f_x, f_y, f_z have been chosen randomly from a Gaussian distribution of width 1.0 centered around zero.

Note, that in the presence of noise, in contrary to the previous simulations, also those conformational motions are allowed that correspond to uphill dynamics in the elastic energy landscape. Furthermore, we need to reconsider the conditions for interactions between the machine and the ligand particle, namely the binding process of the substrate particle and its release after conversion into the product ligand. In the real biological situation with fluctuations present, binding of a ligand to the motor protein should depend on the local conditions inside the binding site. Furthermore, such processes should depend on the concentration of ligands.

In our simplified description of machine operation, with effects of noise being neglected, the substrate binding was always assumed to take place in the equilibrium conformation of the network whereas the particle conversion and product release were carried out in the conformation that corresponded to the equilibrium of the network-substrate complex. In the presence of noise, however, a definite equilibrium conformation of the network may not be reached.

For the investigation including noise, we assume that binding of the substrate ligand to the machine is determined in terms of the local configuration of the previously defined binding pocket. The binding process itself will be considered of stochastic nature, meaning that in the simulation it takes place with a certain rate. One further aspect was involved in our simulations. In any enzyme, the formation of the enzyme-substrate complex upon substrate binding is in principle reversible. This means that the substrate ligand before its conversion into the product can also undergo dissociation from the enzyme with a certain rate.

We have now taken into account this possibility. When the machine-substrate complex became formed and subsequent conformational changes may have taken place

6.5. A PROTOTYPE SYNTHETIC PROTEIN MOTOR

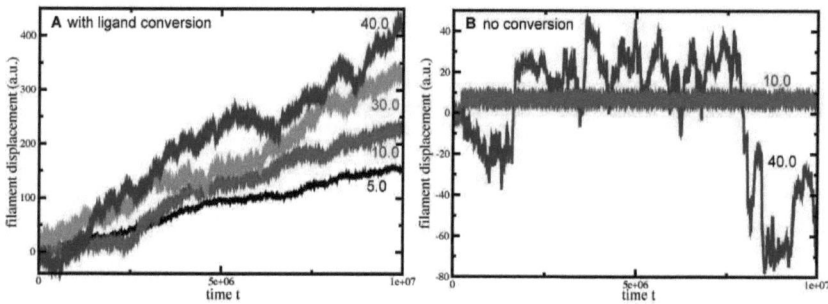

Figure 6.8: **Filament transport by the machine.** The filament displacement is plotted against the time for various values of the noise intensity. (A) In this set of simulations conversion of the ligand from the substrate into the product particle was incorporated. (B) Here, no conversion was implemented and thus only equilibrium fluctuations of the system were examined. The machine activity was based on substrate binding and its dissociation only. In the case of relatively low noise (10.0, red curve) the machine-filament complex with the substrate bound has reached its equilibrium state and no transport is possible. For larger noise (40.0, blue curve) links between the machine and the filament become regularly broken and newly established. However, no net transport of the filament can be achieved.

already, the substrate ligand could still dissociate from the machine with a certain rate, again depending on the conditions inside the binding pocket. The conversion of the substrate into the product particle and its following ejection from the machine were also formulated in terms of local conditions of the interaction pocket.

6.5.3 Filament transport by the motor

To investigate filament transport by the synthetic machine under its ligand-induced operation and in the presence of noise, we have run simulations implementing different dynamical scenarios.

Two major situations were distinguished by us. In a first series of simulations we have focused on the working cycles of the synthetic machine which would emulate the operation of a molecular motor under equilibrium conditions. This means that only processes of substrate binding and dissociation have been taken into account into the dynamical modeling, whereas the reaction and release of products were absent. In a second set of simulations, we have allowed the machine-filament system to perform

CHAPTER 6. MODELS OF SYNTHETIC PROTEIN MOTORS

its operation also beyond equilibrium by including processes of ligand conversion and product release. For both situations relatively large time series of the machine operation have been recorded for different noise intensities. The intensities correspond to different system temperatures.

As mentioned before, the interaction between the machine and the ligands should be described in terms of local conditions of the ligand interaction pocket. To do this, local geometric conditions of the binding pocket have been employed. As an appropriate measure for the size of the pocket we have used the radius of gyration which is calculated as the root mean square distance of particles of the pocket from its geometric center. Its square reads as

$$R_g^2 = (1/B) \sum_{i \in N_p} (\vec{R}_i - \vec{R}_c)^2, \tag{6.2}$$

where the summation is carried out over the B particles that belong to the neighborhood N_p of the binding pocket. They are the three selected particles the ligand can bind to plus the particles that are connected to them. The position vectors of these particles are $\vec{R}_i$, and $\vec{R}_c$ is the position vector of the geometric center of the three binding particles.

As estimated from the previous simulation of the machine operation without the filament (see Fig. 6.6), we have set the threshold for binding of the substrate to $R_g > 5.45$, which means that the machine is still around its structural global equilibrium conformation. If this conditions was fulfilled, binding of the substrate ligand to the actual network structure was possible with the probability of 10^{-4} in each integration step. Once the substrate-network complex has been formed, the substrate was allowed to dissociate under the same conditions under which it became bound to, i.e. assuming the identical probability and the same threshold for R_g.

The ligand conversion and release were allowed to take place with a probability of 10^{-4} in each integration step, provided that local conditions of the active pocket, namely that $R_g < 5.39$, were fulfilled. The conditions controlling the binding of the machine to the filament were the same as in the noiseless case.

We have compared the outcome of a series of simulations showing obvious differences between the operation of the machine operating under equilibrium conditions, i.e. without substrate ligand conversion, and under non-equilibrium conditions where substrate conversion into the product was considered.

In Fig. 6.8 we have plotted the filament displacement under consequent operation of the machine for various values of noise intensity. We can clearly see that steady directional filament transport is possible only with the machine operating under non-equilibrium conditions. When working instead at equilibrium, the machine can temporarily transport the filament in both directions, but, if a long-time average is performed, its net locomotion vanishes and directed transport of the filament is not observed. These results also suggest that in the case of non-equilibrium operation of the synthetic motor, the presence of noise may be beneficial as it increases the transport rate of the filament (see Fig. 6.8).

6.6 Coordinates of the designed network

The following table provides the spatial coordinates of equilibrium position of the 50 particles that constitute the designed network which has been used by us to construct the synthetic motor template. The pattern of spring connections used in our simulations can be reproduced from this data by applying an interaction radius of $l_{int} = 7.5$.

index i	x_i	y_i	z_i
0	0.000000	0.000000	0.000000
1	-2.250606	1.513577	1.814874
2	-3.547681	-0.657242	3.959800
3	-1.694768	-2.842780	5.264394
4	1.494905	-2.547102	3.813177
5	4.529039	-2.770177	1.992979
6	4.072661	0.262305	2.865932
7	1.670645	0.378216	5.765140
8	0.164961	3.089509	7.921995
9	-0.718121	-0.506017	9.003187
10	0.027955	-2.616314	11.803884

CHAPTER 6. MODELS OF SYNTHETIC PROTEIN MOTORS

11	-1.976225	-4.102386	8.773245
12	-5.734028	-3.993969	9.534685
13	-4.059582	-3.257490	12.571988
14	-6.131467	-0.130246	12.892877
15	-9.404208	0.492210	14.177999
16	-11.335046	0.712168	11.093718
17	-8.102510	1.170035	9.877696
18	-10.233940	-0.733887	7.333117
19	-11.294299	-3.519827	8.765336
20	-14.091883	-1.657092	8.496408
21	-15.225877	-2.475998	11.339919
22	-16.454146	-1.633076	15.007511
23	-18.705472	-0.643274	12.435505
24	-21.657510	0.958647	11.536650
25	-22.165406	-0.737916	14.620891
26	-24.366702	-3.826345	15.289950
27	-21.318742	-2.968613	17.208187
28	-20.928556	-4.337320	14.197824
29	-17.618755	-6.479827	14.083390
30	-19.579717	-8.918729	16.121652
31	-17.822841	-7.376985	18.560644
32	-19.004346	-10.464905	19.278277
33	-21.894159	-10.676693	17.450503
34	-22.041187	-8.958193	20.431335
35	-23.935834	-7.863389	18.150190
36	-26.453200	-9.574022	18.060080
37	-28.115449	-6.331969	17.401296
38	-28.889437	-8.304478	14.157759
39	-27.137244	-11.459018	14.761705
40	-23.332556	-10.891567	14.437410
41	-20.822941	-9.175062	12.747098

6.6. COORDINATES OF THE DESIGNED NETWORK

42	-21.989900	-10.823980	9.437601
43	-24.325875	-12.599568	10.867699
44	-22.562342	-14.152048	8.744179
45	-19.851852	-13.165531	10.231882
46	-16.896938	-13.586741	10.974879
47	-13.817907	-15.520423	11.899851
48	-13.617313	-12.161436	13.840984
49	-14.184332	-8.596932	14.647480

Chapter 7

Conclusions

The work presented in this thesis can be divided into two parts. In the main part, we have studied the dynamics of real biological motor proteins whereas in the second part model systems that were analogs of biological objects have been considered.

In the first chapter, we have provided a brief introduction to the field of protein research. We have outlined the importance of enzymes and motor proteins and given a short survey on experimental techniques used to investigate functional aspects of their operation.

In Chapter 2, an approximate description for the dynamics of proteins, viewing them as deformable elastic objects, was introduced. The focus in this chapter was placed on the presentation and discussion of the relaxational elastic network model, which was the particular model used in this thesis. Its advantage is that it allowed us to study large-amplitude slow protein motions with structural resolution.

A large part of the work was devoted to the investigation of the hepatitis C virus (HCV) helicase motor using this coarse-grained model. At the first stage, we have probed dynamical properties of this protein. We have shown that coordinated domain motions are underlying its organized dynamics. They are used to execute robust operation cycles that are related to ATP binding and hydrolysis. We could reproduce the conformational motions accompanying the ATP-dependent operation cycle of this helicase motor by incorporating interactions with ATP into the model. These results are described in the third chapter.

In Chapter 4, a dynamical description for the DNA molecule has been implemented and interactions between the helicase and DNA have been included. With such model extension, we could trace entire operation cycles of the HCV helicase in a structurally resolved way. As we have shown, the motor can move base-by-base along the DNA consuming one ATP per travelled base. Under this translocation process one protein domain was acting as a mechanical wedge separating the duplex DNA.

Previously, the *ratcheting inchworm* translocation mechanism has been proposed based on the experimental data. The results of our theoretical modeling confirm this mode of operation. Recently obtained crystallographic data of HCV helicase complexes agrees well with our findings [137, 138]. The single-base pair unwinding of HCV helicase as predicted by our computer simulations has been also confirmed recently in single-molecule experiments [146].

Chapter 5 is devoted to the study of three other helicase proteins of the largest superfamily. We have probed their mechanical properties and analyzed large-amplitude conformational motions using similar methods as for the HCV helicase. The common observed property for all investigated helicases is that they are able to perform robust and well-coordinated domain motions that are underlying their ATP-dependent activity. On the other hand, we have also observed differences in the conformational dynamics which can be important for the biological operation of the helicases. The studied helicases are poorly explored in experiments yet. Our predicted conformational changes can be tested in further experiments.

The dynamical systems investigated in Chapter 6 are of different nature. There, we have considered artificially constructed analogs of real biological objects. The construction of a prototype synthetic protein motor operating on a filament has been explained and its behavior in the presence of thermal fluctuations investigated. We have shown that steady directional filament transport by this model motor is possible only when operating under non-equilibrium conditions.

In our study, the operation of a real molecular motor could be reproduced - for the first time - in structurally resolved dynamical simulations. In general, our results are a demonstration of the feasibility of coarse-grained modeling, bridging the gap between full molecular dynamics and reduced phenomenological theories of molecular motors.

Any coarse-grained models, including elastic-network descriptions, are intrinsically

CHAPTER 7. CONCLUSIONS

limited in their explanatory power and cannot fully describe molecular operation mechanisms of motor proteins. While chemical details are not resolved in elastic network models, it is, however, remarkable that such purely mechanical descriptions are still able to capture essential aspects of functional dynamics in protein motors. Apparently, the operation of molecular motors is so robust that, in a rough approximation, it is not sensitive to fine chemical details.

In the future it can be however interesting to develop and apply hybrid descriptions, combining coarse-grained elastic network models with more realistic molecular dynamics simulations.

Our investigations performed for a prototype synthetic motor have demonstrated the advantage of such replacements of real motors in theoretical modeling. These models allow to study the operation of molecular motors in a structurally resolved fashion under controllable conditions. In forthcoming studies we plan to extend the present model systems further to incorporate also aspects of reversibility of motor operation. Moreover, we want to consider its activity to be controlled by the concentration of substrate and product particles.

The method of evolutionary optimization used by us to design special elastic network templates with robust dynamical properties can be also varied. Interesting scenarios can actually be considered: Today we know that different motor proteins have originated from gene duplication events and thus have common ancestors. Under biological evolution, subject to various evolutionary pressure, their substrate-dependent activity has however evolved in different directions. As a consequence, their global structures are different and also their functional properties.

One can therefore consider design processes of artificial networks in which evolution would be performed with respect to the ligand-induced activity. Then, optimization can be, for instance, aimed to design networks with tailored types of ligand-induced conformational motions. In this situation one would be able to investigate the impact of various types of principal motions on, e.g. the efficiency of a molecular motor.

The operation of many real molecular machines relies on the coordinated interaction between multiple structural subunits. The ligand-powered artificial machine designed by us can be further used to construct a more complicated model system consisting of several coupled units. Such a system can mimic oligomeric molecular machines and may serve as an appropriate model to study the effects of cooperativity and communication in such biological devices.

Bibliography

[1] Sewald N, Jakubke H D (2009) Peptides: Chemistry and biology. *Wiley-VCH, 2nd revised and updated Edition*

[2] Whitford D (2005) Proteins: Structure and function. *John Wiley & Sons, 1st Ed.*

[3] Alberts B, et al. (2002) Molecular biology of the cell. *Garland Science, New York, 4th Ed.*

[4] Nelson D L, Cox M M (2008) Lehninger, Principles of biochemistry. *Palgrave Macmillan, 5th Ed.*

[5] Patthy L (2007) Protein evolution. *Wiley-Blackwell, 2nd Ed.*

[6] Eigen M (1971) Selforganization of matter and the evolution of biological macromolecules. *Naturwissenschaften* 58(10):465-523

[7] Tokuriki N, Tawfik D S (2009) Protein dynamism and evolvability. *Science* 324:203-207

[8] Griffiths A J F, Gelbart W M, Miller J H, Lewontin R C (1999) Modern genetic analysis. *W H Freeman, New York*

[9] Bischoff T L W, Voit C (1860) Die Gesetze der Ernährung des Pflanzenfressers durch neue Untersuchungen festgestellt. *Leipzig, Heidelberg*

[10] Sanger F, Tuppy H (1951) The amino-acid sequence in the phenylalanyl chain of insulin. 1. The identification of lower peptides from partial hydrolysates. *Biochem J* 49:463-481

[11] Sanger F, Tuppy H (1951) The amino-acid sequence in the phenylalanyl chain of insulin. 2. The investigation of peptides from enzymic hydrolysates. *Biochem J* 49:481-490

[12] Kendrew J et al (1958) A three-dimensional model of the myoglobin molecule obtained by X-ray analysis. *Nature* 181:662-666

[13] Muirhead H, Perutz M (1963) Structure of hemoglobin. A three-dimensional fourier synthesis of reduced human hemoglobin at 5.5 Å resolution. *Nature* 199:633-638

[14] Voet D J, Voet J G (2004) Biochemistry. *John Wiley & Sons, 3rd Ed.*

[15] Selvin P R, Ha T, eds. (2007) Single molecule techniques: A laboratory manual. *Cold Spring Harbor Laboratory, 1st Ed.*

[16] Yanagida T, Ishii Y, eds. (2008) Single molecule dynamics in life science. *Wiley-VCH, 1st Ed.*

[17] Howard J (2001) Mechanics of motor proteins & the cytoskeleton. *Palgrave Macmillan*

[18] Alberts B (1998) The cell as a collection of protein machines: Preparing the next generation of molecular biologists. *Cell* 92:291-294

[19] Juelicher F, Ajdari A, Prost J (1997) Modeling molecular motors. *Rev Mod Phys* 69:1269-1282

[20] Astumian R D (1997) Thermodynamics and kinetics of a brownian motor. *Science* 276:917-922

[21] Cui Q, Bahar I, eds. (2006) Normal mode analysis: Theory and Application to biological and chemical systems. *CRC, Boca Raton, FL*

[22] Tirion M M (1996) Large-amplitude elastic motions in proteins from a single-parameter, atomic analysis. *Phys Rev Lett* 77:1905

[23] Ma J, ed. (2006) Gene expression and regulation. *Springer, 1st Ed.*

BIBLIOGRAPHY

[24] Anfinsen C B (1973) Principles that govern the folding of protein chains. *Science* 181:223-230

[25] Creighton T E (1993) Proteins: Structures and molecular properties. *W H Freeman and Company, New York, 2nd Ed.*

[26] Bukau B, Deuerling E, Pfund C, Craig E A (2000) Getting newly synthesized proteins into shape. *Cell* 101:119-122

[27] Baldwin R L (1994) Matching speed and stability. *Nature* 369:183-184

[28] Daggett V, Fersht A R (2003) Is there a unifying mechanism for protein folding? *Trends Biochem Sci* 28:18-25

[29] Rose G D, Fleming P J, Banavar J R, Maritan A (2006) A backbone-based theory of protein folding. *Proc Natl Acad Sci U S A* 103:623-633

[30] Shakhnovich E I, Gutin A M (1993) Engineering of stable and fast-folding sequences of model proteins. *Proc Natl Acad Sci USA* 90:7195-7199

[31] Leopold P E, Montal M, Onuchic J N (1992) Protein folding funnels: A kinetic approach to the sequence-structure relationship. *Proc Natl Acad Sci USA* 89:8721-8725

[32] Onuchic J N, Wolynes P G (2004) Theory of protein folding. *Curr Opin Struct Biol* 14:70-75

[33] Soskine M, Tawfik D S (2010) Mutational effects and the evolution of new protein functions. *Nature Reviews Genetics* 11:572-582

[34] Friedmann H, ed. (1981) Benchmark papers in biochemistry, Vol. 1: Enzymes. *Hutchinson Ross Publishing Company, Stroudsburg, PA*

[35] Kornberg A (1987) For the love of enzymes: The odyssey of a biochemist. *Harvard University Press, Cambridge*

[36] Kraut J (1988) How do enzymes work? *Science* 242:533-540

[37] Jencks W P (1987) Catalysis in chemistry and enzymology. *Dover Publications, Inc, New York*

[38] Fersht A (1999) Structure and mechanism in protein science: A guide to enzyme catalysis and protein folding. *W H Freemann and Company, New York*

[39] Gutteridge A, Thornton J M (2005) Understanding nature's catalytic toolkit. *Trends Biochem Sci* 30:622-629

[40] Menten L, Michaelis M I (1913) Die Kinetik der Inertinwirkung. *Biochem Z* 49:333-369

[41] Cleland W W (1977) Determining the mechanisms of enzyme-catalyzed reactions by kinetic studies. *Adv Enzymol* 45:273-387

[42] Vale R D (2003) The molecular motor toolbox for intracellular transport. *Cell* 112:467-480

[43] Spudich J A (1994) How molecular motors work. *Nature* 372:515-518

[44] Vale R D, Milligan R A (2000) The way things move: Looking under the hood of molecular motor proteins. *Science* 288:88-95

[45] Molloy J E, Veigel C (2003) Myosin motors walk the walk. *Science* 300:2045-2046

[46] Deuerling E, Bukau B (2004) Chaperone-assisted folding of newly synthesized proteins in the cytosol. *Crit Rev Biochem Mol Biol* 39:261-277

[47] Geeves M A, Holmes K C (1999) Structural mechanism of muscle contraction. *Annu Rev Biochem* 68:687-728

[48] Gouaux E, MacKinnon R (2005) Principles of selective ion transport in channels and pumps. *Science* 310:1461-1465

[49] Stokes D L, Green N M (2003) Structure and function of the calcium pump. *Annu Rev Biophys Biomol Struct* 32:445-468

BIBLIOGRAPHY

[50] Caruthers J M, McKay D B (2002) Helicase structure and mechanism. *Curr Opin Struct Biol* 12:123-133

[51] Enemark E J, Joshua-Tor L (2008) On helicases and other motor proteins. *Curr Opin Struct Biol* 18:243-257

[52] Lohman T M, Bjornson K P (1996) Mechanisms of helicase-catalyzed DNA unwinding. *Annu Rev Biochem* 65:169-214

[53] Jankowsky E, Fairman M E (2007) RNA helicases - one fold many functions. *Curr Opin Struct Biol* 17:316-324

[54] Frick D N, Lam A M I (2006) Understanding helicases as a means of virus control. *Curr Pharm Des* 12:1315-1338

[55] Kwong A D, Govinda Rao B, Jeang K T (2005) Viral and cellular RNA helicases as antiviral targets. *Nat Rev Drug Discovery* 4:845-853

[56] Lee J Y, Yang W (2006) UvrD helicase unwinds DNA one base at a time by a two-part power stroke. *Cell* 127:1349-1360

[57] Enemark E J, Joshua-Tor L (2006) Mechanism of DNA translocation in a replicative hexameric helicase. *Nature* 442:270-275

[58] Cann A J (2001) Principles of molecular virology. *Academic Press, 3rd Ed.*

[59] Singleton M R, Dillingham M S, Wigley D B (2007) Structure and mechanism of helicases and nucleic acid translocases. *Annu Rev Biochem* 76:23-50

[60] Pyle A M (2008) Translocation and unwinding mechanism of RNA and DNA helicases. *Annu Rev Biophys* 37:317-336

[61] Lohman T M, Tomko E J, Wu C G (2008) Non-hexameric DNA helicases and translocases: mechanisms and regulation. *Nat Rev Mol Cell Biol* 9:391-401

[62] Gorbalenya A E, Koonin E V (1993) Helicases: Amino acid sequence comparisons and structure-function relationships. *Curr Opin Struct Biol* 3:419-429

[63] Patel S S, Picha K M (2000) Structure and function of hexameric helicases. *Annu Rev Biochem* 69:651-697

[64] Fairman-Williams M E, Guenther U P, Jankowsky E (2010) SF1 and SF2 helicases: family matters. *Curr Opin Struct Biol* 20:313-324

[65] Lattman E, Loll P J, Loll P (2008) Protein crystallography: a concise guide. *JHU Press*

[66] Drenth J (2010) Principles of protein x-ray crystallography. *Springer Verlag*

[67] Rule G S, Hitchens T K (2006) Fundamentals of protein NMR spectroscopy. *Springer, Focus on Structural Biology, Vol 5*

[68] Cavanagh J, Fairbrother W J, Palmer III A G, Skelton N J, Rance M (2007) Protein NMR spectroscopy: principles and practice. *Academic Press, Boston, 2nd Ed.*

[69] Gerstein M, Krebs W (1998) A database of macromolecular motions. *Nucleic Acids Res* 26:4280-4290

[70] Krebs W G, Gerstein M (2000) The morph server: a standardized system for analyzing and visualizing macromolecular motions in a database framework. *Nucleic Acids Res* 28:1665-1675

[71] Echols N, Milburn D, Gerstein M (2003) MolMovDB: analysis and visualization of conformational change and structural flexibility. *Nucleic Acids Res* 31:478-482

[72] Giepmans B N G, Adams S R, Ellisman M H, Tsien R Y (2006) The fluorescent toolbox for assessing protein location and function. *Science* 312:217-224

[73] Joo C, Ishitsuka Y, Buranachai C, Ha T (2008) Advances in single-molecule fluorescence methods for molecular biology. *Annu Rev Biochem* 77: 51-76

[74] Ha T, et al. (1996) Probing the interaction between two single molecules: fluorescent resonance energy transfer between a single donor and a single acceptor. *Proc Natl Acad Sci U S A* 93:6264-6268

[75] Rahul R, Sungchul H, Ha T (2008) A practical guide to single-molecule FRET. *Nature Methods* 5:507-516

[76] Ha T (2001) Single-molecule fluorescence resonant energy transfer. *Methods* 25:78-86

[77] Bustamante C, Bryant Z, Smith S B (2003) Ten years of tension: single-molecule DNA mechanics. *Nature* 421:423-427

[78] Forde N R, Izhaky D, Woodcock G R, Wuite G J, Bustamante C (2002) Using mechanical force to probe the mechanics of pausing and arrest during continuous elongation by Escherichia coli RNA polymerase. *Proc Natl Acad Sci U S A* 99:11682-11687

[79] Kellermayer M S Z, Bustamante C, Granzier H L (2003) Mechanics and structure of titin oligomers explored with atomic force microscopy. *Biochimica Et Biophysica Acta - Bioenergetics* 1604:105-114

[80] Kellermayer M S, Smith S, Bustamante C, Granzier H L (2000) Mechanical manipulation of single titin molecules with laser tweezers. *Adv Exp Med Biol* 481:111-126

[81] Giessibl F J (2003) Advances in atomic force microscopy. *Rev Mod Phys* 75:949

[82] Butt H, Cappella B, Kappl M (2005) Force measurement with the atomic force microscope: technique, interpretation and application. *Surface Science Reports* 59:1-152

[83] Kodera N, Yamamoto D, Ishikawa R, Ando T (2010) Video imaging of walking myosin V by high-speed atomic force microscopy. *Nature* 468:72-76

[84] Uchihashi T, Iino R, Ando T, Noji H (2011) High-speed atomic force microscopy reveals rotary catalysis of rotorless F1-ATPase. *Science* 333:755-758

[85] McCammon J A, Gelin B R, Karplus M (1977) Dynamics of folded proteins. *Nature* 267:585-590

[86] Weiner P W, Kollman P A (1981) AMBER: assisted model building with energy refinement. A general program for modeling molecules and their transitions. *J Comput Chem* 2:287-303

[87] Brooks B R, et al. (1983) CHARMM: a program for macromolecular energy, minimization, and dynamical calculations. *J Comput Chem* 4:187-217

[88] Scott W R P, et al. (1999) The GROMOS biomolecular simulation program package. *J Phys Chem A* 103:3596-3607

[89] van Gunsteren W F, Weiner P K, Wilkinson A J (1993) Computational simulation of biomolecular systems: theoretical and experimental applications. Vol 2. *ESCOM, Leiden*

[90] Tuckerman M E, Martyna G J (2000) Understanding modern molecular dynamics: techniques and applications. *J Phys Chem* 104:159-178

[91] Karplus M, McCammon J A (2002) Molecular dynamics simulations of biomolecules. *Nat Struct Biol* 9:646-652

[92] Shaw D E (2010) Atomic-level characterization of the structural dynamics of proteins. *Science* 330:341-346

[93] Gaspard P, Gerritsma E (2007) The stochastic chemomechanics of the F1-ATPase molecular motor. *J Theor Biol* 247:672-686

[94] Howard J (2009) Motor proteins as nanomachines: the roles of thermal fluctuations in generating force and motion. *Seminaire Poincare XII* 33-44

[95] Astumian R D, Haenggi P (2002) Brownian motors. *Physics Today* 55:33-39

[96] Haenggi P, Marchesoni F, Nori F (2005) Brownian motors. *Ann Phys, Leipzig* 14:51-70

[97] Hinsen K, Petrescu A J, Dellerue S, Bellissent-Funel M C, Kneller G R (2000) Harmonicity in slow protein dynamics. *Chemical Physics* 261:25-37

[98] Tozzini V (2005) Coarse-grained models for proteins. *Curr Opin Struct Biol* 15:144-150

[99] Tozzini V (2009) Multiscale modeling of proteins. *Accounts of Chemical research* 43:220-230

[100] Ueda Y, Taketomi H, Go N (1978) Studies on protein folding, unfolding and fluctuations by computer simulation. A three-dimensional lattice model of lysozyme. *Biopolymers* 17:1531-1548

[101] Hills Jr R D, Brooks III C L (2009) Insights from coarse-grained Go models for protein folding and dynamics. *Int J Mol Sci* 10:889-905

[102] Bahar I, Atilgan A R, Erman B (1997) Direct evaluation of thermal fluctuations in proteins using s single-parameter harmonic potential. *Fold Des* 2:173-181

[103] Hinsen K (1998) Analysis of domain motions by approximate normal mode calculations. *Proteins* 33:417-429

[104] Haliloglu T, Bahar I, Erman B (1997) Gaussian dynamics of folded proteins. *Phys Rev Lett* 79:3090-3093

[105] Bahar I, Atilgan A R, Demirel M C, Erman B (1998) Vibrational dynamics of folded proteins: significance of slow and fast motions in relation to function and stability. *Phys Rev Lett* 80:2733-2736

[106] Atilgan A R et al. (2001) Anisotropy of fluctuation dynamics of proteins with an elastic network model. *Biophys J* 80:505-515

[107] Harrison R W (1984) Variational calculation of the normal modes of a large macromolecule: methods and some initial results. *Biopolymers* 23:2943-2949

[108] Brooks B, Karplus M (1985) Normal modes for specific motions of macromolecules: application to the hinge-bending mode of lysozyme. *Proc Natl Acad Sci USA* 82:4995-4999

[109] Case D A (1994) Normal mode analysis of protein dynamics. *Curr Opin Struct Biol* 4:285-290

[110] Marques O, Sanejouand Y H (1995) Hinge-bending motion in citrate synthase arising from normal mode calculations. *Proteins* 23:557-560

[111] Tama F, Sanejouand Y H (2001) Conformational change of proteins arising from normal mode calculations. *Protein Engineering* 14:1-6

[112] Zheng W, Doniach S (2003) A comparative study of motor-protein motions by using a simple elastic-network model. *Proc Natl Acad Sci U S A* 100:13253-13258

[113] Yang L, Song G, Jernigan R L (2007) How well can we understand large-scale protein motions using normal modes of elastic neworks. *Biophys J* 93:920-929

[114] Purcell E M (1977) Life at low reynolds numbers. *American Journal of Physics* 45:3-11

[115] Kitao A, Hirata F, Go N (1991) The effects of solvent on the conformation and the collective motion of proteins: Normal mode analysis and molecular dynamics simulations of melittin in water and vacuum. *Chem Phys* 158:447-472.

[116] Dietz H, Rief M (2008) Elastic bond network model for protein unfolding mechanics. *Phys Rev Lett* 100:098101

[117] Echevarria C, Togashi Y, Mikhailov A S, Kapral R (2011) A mesoscopic model for protein enzymatic dynamics in solution. *Phys Chem Chem Phys* 13:10527-10537

[118] Togashi Y, Mikhailov A S (2007) Nonlinear relaxation dynamics in elastic networks and design principles of molecular machines. *Proc Natl Acad Sci U S A* 104:8697-8702

[119] Togashi Y, Yanagida T, Mikhailov A S (2010) Nonlinearity of mechanochemical motions in motor proteins. *PLoS Comput Biol* 6(6):e1000814

[120] Cressman A, Togashi Y, Mikhailov A S, Kapral R (2008) Mesoscale modeling of molecular machines: Cyclic dynamics and hydrodynamical fluctuations. *Phys Rev E* 77:050901(R)

[121] Yang L, Song G, Jernigan R L (2009) Protein elastic network models and the ranges of cooperativity. *Proc Natl Acad Sci USA* 106:12347-12352

BIBLIOGRAPHY

[122] Hinsen K, Kneller GR (1999) A simplified force field for describing vibrational protein dynamics over the whole frequency range. *J Chem Phys* 111:10766-10769

[123] Hinsen K (2008) Structural flexibility in proteins: Impact of the crystal environment. *Bioinformatics* 24:521-528

[124] Hinsen K (2009) Physical arguments for distance-weighted interactions in elastic network models for proteins. *Proc Natl Acad Sci USA* 106:E128

[125] Lin C P, et al (2008) Deriving protein dynamical properties from weighted protein contact number. *Proteins* 72:929-935

[126] Riccardi D, Cui Q, Phillips Jr. G N (2010) Evaluating elastic network models of crystalline biological molecules with temperature factors, correlated motions, and diffusive x-ray scattering. *Biophys J* 99:2616-2625

[127] Parker D, Bryant Z, Delp S L (2009) Coarse-grained structural modeling of molecular motors using multibody dynamics. *Cellular and Molecular Bioengineering* 2:366-374

[128] Stember J N, Wriggers W (2009) Bend-twist-stretch model for coarse elastic network simulations. *J Chem Phys* 131:074112

[129] Mendez R, Bastolla U (2010) Torsional network model: Normal modes in torsion angle space better correlate with conformation changes in proteins. *Phys Rev Lett* 104:228103

[130] Tan S-L, ed. (2006) Hepatitis C virus, Genomes and molecular biology. *Horizon Bioscience, Norfolk (UK)*

[131] Tang H, Grise H (2009) Cellular and molecular biology of HCV infection and hepatitis. *Clin Sci (London)* 117:49-65

[132] Frick D N (2007) The hepatitis C virus NS3 protein: A model RNA helicase and potential drug target. *Curr Issues Mol Biol* 9:1-20

[133] Yao N, et al. (1997) Structure of the hepatitis C virus RNA helicase domain. *Nat Struct Biol* 4:463-467

[134] Kim J L, et al. (1998) Hepatitis C virus NS3 RNA helicase domain with a bound oligonucleotide: The crystal structure provides insights into the mode of unwinding. *Structure* 6:89-100

[135] Dumont S, et al. (2006) RNA translocation and unwinding mechanism of HCV NS3 helicase and its coordination by ATP. *Nature* 439:105-108

[136] Myong S, Bruno M C, Pyle A M, Ha T (2007) Spring-loaded mechanism of DNA unwinding by hepatitis C virus NS3 helicase. *Science* 317:513-516

[137] Gu M, Rice C M (2010) Three conformational snapshots of the hepatitis C virus NS3 helicase reveal a ratchet translocation mechanism. *Proc Natl Acad Sci U S A* 107:521-528

[138] Appleby, et al. (2011) Visualizing ATP-dependent RNA translocation by the NS3 helicase from HCV. *J Mol Biol* 405:1139-1153

[139] Flechsig H, Mikhailov A S (2010) Tracing entire operation cycles of molecular motor hepatitis C virus helicase in structurally resolved dynamical simulations. *Proc Natl Acad Sci U S A* 107:20875-20880

[140] Frick D N, Rympa R S, Lam A M I, Frenz C M (2004) Electrostatic analysis of the hepatitis C virus NS3 helicase reveals both active and allosteric site locations. *Nucleic Acids Res* 32:5519-5528

[141] Cocco S, Monasson R, Marko J F (2001) Force and kinetic barriers to unzipping of the DNA double helix. *Proc Natl Acad Sci U S A* 98:8608-8613.

[142] Sakaue T, Yoshikawa K (2001) Folding/unfolding kinetics on a semiflexible polymer chain. *J Chemp Phys* 117:6323-6330

[143] Kierfeld J (2006) Force-induced desorption and unzipping of semi-flexible polymers. *Phys Rev Lett* 97:058302

[144] Chakrabarti B, Nelson D R (2009) Shear unzipping of DNA. *J Phys Chem B* 113:3831-3836

[145] Myong S, Ha T (2010) Stepwise translocation of nucleic acid motors. *Curr Opin Struct Biol* 20:121-127

[146] Cheng W, Arunajadai S G, Moffitt J R, Tinoco Jr I, Bustamante C (2011) Singel-base pair unwinding and asynchronous RNA release by the hepatitis C virus NS3 helicase. *Science* 333:1746-1749

[147] Flechsig H, Popp D, Mikhailov A S (2011) In silico investigation of conformational motions in superfamily 2 helicase proteins. *PLoS ONE* 6(7):e21809

[148] Nishino T, Komori K, Tsuchiya D, Ishino Y, Morikawa K (2005) Crystal structure and functional implications of Pyrococcus furiosus hef helicase domain involved in branched DNA processing. *Structure* 13:143-153

[149] Buettner K, Nehring S, Hopfner K P (2007) Structural basis for DNA duplex separation by a superfamily-2 helicase. *Nat Struct Mol Biol* 14:647-652

[150] Liu H, et al. (2008) Structure of the DNA repair helicase XPD. *Cell* 133:801-812

[151] Richards J D, et al. (2008) Structure of the DNA repair helicase Hel308 reveals DNA binding and autoinhibitory domains. *J Biol Chem* 283:5118-5126

[152] Woodman I L, Briggs G S, Bolt E L (2007) Archaeal Hel308 domain V couples DNA binding to ATP hydrolysis and positions DNA for unwinding over the helicase ratchet. *J Mol Biol* 374:1139-1144

[153] Luo D H, et al. (2008) Insights into RNA unwinding and ATP hydrolysis by the flavivirus NS3 protein. *EMBO Journal* 27:3209-3219

[154] Soultanas P, Wigley D B (2001) Unwinding the Gordian knot of helicase action. *Trends Biochem Sci* 26:47-54

i want morebooks!

Buy your books fast and straightforward online - at one of world's fastest growing online book stores! Environmentally sound due to Print-on-Demand technologies.

Buy your books online at
www.get-morebooks.com

Kaufen Sie Ihre Bücher schnell und unkompliziert online – auf einer der am schnellsten wachsenden Buchhandelsplattformen weltweit! Dank Print-On-Demand umwelt- und ressourcenschonend produziert.

Bücher schneller online kaufen
www.morebooks.de

VDM Verlagsservicegesellschaft mbH
Heinrich-Böcking-Str. 6-8
D - 66121 Saarbrücken

Telefon: +49 681 3720 174
Telefax: +49 681 3720 1749

info@vdm-vsg.de
www.vdm-vsg.de

Printed by Books on Demand GmbH, Norderstedt / Germany